I0470884

AMERICA'S SUPER SECRET AIRCRAFT
And
The Deadly Drones

Breaking the sound barrier

by Bonnie Henderson
Editor: John Tomikel

ISBN 978 1490553306
also 1490553304

(There is a Timeline followed by an addendum of terms, people, and places at the end of this work)

Editor Note: The author had worked on her UFO book for eight years. She came to the conclusion that the UFO phenomena was the result of the development of great fighting planes shortly after World War II and this development continues to the present day. Most of the super secret aircraft were reported as UFOs after their maiden flights. Therefore, the subject is approached from the viewpoint that the spectacular flights reported as UFOs were test runs for the super aircraft. The UFOs were clues that led to the discovery of classified aircraft and their super abilities.

Prologue

The Projected Secret Aircraft of the Future
This introduction is based on an article that appeared in Popular Science, June 2015 and written by Clay Dillow

No one able to read these pages needs to be reminded of the necessity of America and its Allies to have superiority in the air and the ability to see what our potential enemies have developed. Thus, there are aircraft developed for combat and aircraft developed for spying. Drone and satellite photography are just two of the tools in our American warehouse. It is essential to have something, perhaps known to our potential enemies, but not available to them. It is essential to spy on them just as it is essential for them to spy on us.

The greatest development in the aviation past in this area was the creation of SR-71 Blackbird, which at the beginning of the twenty-first century was (and is) the fastest air-breathing military aircraft the world has ever known. It still flies so high enemy defenses can not intercept it. Satellite technology was able to reduce its advantage. As in all military maneuvers it is essential to keep ahead of the competition. And also, the development of ground to air missiles has reduced the capacity of air-breathing aircraft to be completely effective.

Aerospace engineers believe we can create a super drone aircraft that will fly at four thousand miles an hour. The drone has the tentative name SR 72. This aircraft is expected to attack targets with speeds up to Mach 8. The SR 71 can reach Mach 4 on a good day.

Aeronautical engineers at Lockheed Martin and Aeroject Rocketdyne have been designing the SR-72 at the Skunk Works black site in California. The Skunk Works is mentioned elsewhere in this text concerning its former secret projects.

The new design will use a hybrid propulsion system and a conventional turbo-jet and be able to take-off at the speed of Mach 3. Once it's in the air its power will boost it to higher velocities. The materials of construction will have to be so engineered that it can

3

withstand temperatures of speeds that are high enough to melt steel.

The speed of the aircraft will be so great it could reach any location on earth in less than an hour. You won't hear or see it coming.

Where are we today with the essentials of propulsion, skin, frame and payload?

Propulsion – Turbojet planes can get up to Mach 3. In order to get to a higher speed the plane must have a ram-jet which depends on air intake. To make this work the turbo-jet takes the aircraft to Mach 3 and a switch to ram-jet takes it a step further. The final stage is a switch to scramjet mode which uses highly compressed air for combustion, sort of like the old high compression car engine, but with much higher compression.

Skin – Any speed higher than Mach 5 will heat the surface of the plane to above two thousand degrees Fahrenheit. Since conventional frames and skins will melt at that temperature there is a scramble to create new materials that can withstand that phenomena. Materials made from ceramics and carbon compounds have shown great promise for the future and have been quite successful in the past. Any leak in the skin will permit the heat to enter the plane and cause it to disintegrate.

Frame – If you are standing in the back of an open pickup truck and it suddenly stops or accelerates you will be thrown about. Imagine the stress on the structure of a space vehicle when it shifts from one acceleration to the next higher one. Not only is the strength of the frame important, but so is the design that reduces friction and air flow.

Payload – How do you drop a bomb traveling at the speed of Mach 4 or Mach 5 and still hit a target. Even taking photos at that speed is probably impossible. Getting a bead on a target seven or eight miles below and launching a strike on it while traveling two thousand or more miles an hour will take super engineering and super genius.

The SR-71 Blackbird is one of the most spectacular aircraft ever built. It is a long-range, supersonic reconnaissance aircraft capable of flying at Mach 3.2. When it first flew, it was an amazing performer and still is after three decades of unmatched capabilities.

The SR-71 has serviced the United States for more than 35 years. During that time, it has had a very interesting history. It all began back in the mid 1950s when the United States Air Force and the CIA decided that it would be best to replace the U-2, an aircraft with something that would travel much faster and higher to avoid enemy defenses. Lockheed, the developer of the U-2 was also given the contract to develop this supersonic aircraft after a competition with Convair. The project was called ARCHANGEL and the Skunk Works, a division of the Lockheed Aircraft Corporation went through twelve design proposals before they reached their final design, the A-12.

On January 26, 1960, the CIA ordered twelve A-12 aircraft. The next month, Lockheed began to search for 24 pilots for the A-12. Soon after in May of 1960, Francis Gary Powers was shot down in a U-2 over the Soviet Union. This event resulted in the United States and the Soviet Union signing an agreement not to fly manned vehicles over the Soviet Union again, a treaty that was undermined even before the

SR-71 was built.

Various sections of this work appear under other copyrights such as in ISBN 978 0910042 031

Contents In Order of Appearance

Introduction: High Strangeness

INTRODUCTION: HIGH STRANGENESS

If a scientific heresy is ignored or denounced by the general public, there is a chance it may be right. If a scientific heresy is emotionally supported by the general public, it is almost certainly wrong. --Isaac Asimov ("Asimov's Corollary")

On October 18, 1973 a UFO incident occurred that was rated as one of the most credible in the "high strangeness" category events. Around 10:30 PM a UH-IH helicopter of the U.S. Army Reserve left Port Columbus, Ohio for its home base at Cleveland Hopkins airport. Captain Lawrence J. Coyne, age 36, with 19 years of flying experience, was in command. First lieutenant Arrigo Jezzi, a chemical engineer, was at the controls. Also on board was Sergeant John Healey, a Cleveland policeman and flight medic and Crew Chief Sergeant Robert Yanacsek, a computer technician.

The helicopter was cruising at 2500 feet with airspeed of 90. It was a clear, calm night. Near Mansfield, Ohio Sergeant Healey noticed a single red light off to the west that appeared to be heading south. About two minutes later, Sergeant Yanacsek also noticed a single red light on the southeast horizon. He watched it for a while before mentioning it to Captain Coyne who told him to keep an eye on it."

Suddenly Yanacsek announced the red light had changed direction and was now heading toward the helicopter, apparently on a collision course. Coyne grabbed the controls from Jezzi and powered the helicopter into a steep descent at 500 feet per minute. At the same instant, the Captain made radio contact with the Mansfield control tower and then immediately lost it. Jezzi tried to reestablish contact, but without success.

The red light continued coming toward them and its speed seemed to be increasing as much as 700 miles per hour. A collision seemed imminent. "Oh God, this is it!" Coyne thought. He increased the rate of descent to 2000 feet per minute and the airspeed to 100 knots. He was down to 1700 feet and the crew braced for impact. But then the light stopped and took up a hovering position above and in front of the helicopter. "It wasn't cruising, it was stopped. For maybe ten to twelve seconds, just stopped," reported Yanascek.

In front of the windshield was a cigar-shaped, featureless,gray metallic-looking object with a slight dome. "It was shaped fat cigar," Coyne later said, and "it had a big, gray, metallic hull about sixty feet long." Yanacsek thought there was a "a suggestion of windows" along the top of the dome. The red light was coming from the bow. From the slightly indented stem came a white light and from below shone a "green pyramid-shaped" beam. The beam swung upward and flooded the helicopter cockpit with green light. Recalled Coyne, "It was shining brightly through the bubble canopy ... turning everything inside green."

After about ten seconds, the Unidentified Flying Object (UFO) began moving off toward the west, with only the white light visible. The white light maintained its intensity, even as it moved farther away, until it suddenly executed a 45-degree turn to the right, headed out toward Lake Erie and "snapped out" over the horizon.

The object had made no noise and the crew hadn't noticed any turbulence from the encounter except for a "bump" when the object moved away from them to the west. After the UFO moved off, Coyne and Jezzi noticed the magnetic compass disk was rotating about four times a minute. The altimeter read 3500 feet meaning the helicopter was climbing at 1000 feet per minute, although Coyne insisted he had set the controls for descent. It reached nearly 3800 feet before Coyne was able to bring it under control and return to the flight plan altitude of 2500 feet. Only Coyne had been aware of the helicopter's climb, although the other crew members had been acutely aware of the g-forces generated by the dive. The crew reestablished radio contact and the flight proceeded to Cleveland as planned.

People on the ground also witnessed the encounter. Around 11 :00 PM Mrs. E.C. and four children were driving south from Mansfield when they noticed a single, steady, bright, red light flying south. About five minutes later they saw two bright lights, one red and another green, coming rapidly toward them from the southeast. Mrs. C. pulled over to the side of the road. The lights slowed and moved to the car's right. Then another group of lights came at them from the

8

southwest. Some of these lights were flashing and there was "a beating sound, a lot of racket."

Two of the kids jumped from the car and saw the helicopter and the UFO that they described as "like a blimp, as big as a school bus, sort of pear-shaped." At this point, the helicopter was about 650 feet above the trees. The object's green light came on and "it was like rays coming down. The helicopter, the trees, the road, the car, everything turned green."

The "Coyne Helicopter Incident" as it would later be known was investigated by authorities, but no satisfactory explanation was ever presented. Some ufologists speculated the object was a spaceship with some kind of anti-gravity propulsion system that caused the helicopter to rise, despite its controls being set for descent. Coyne for his part stuck to his story although he said he didn't believe in UFOs. Philip Klass, a UFO skeptic, thought the crew had seen a fireball that was part of the Orionid meteor shower that was common at that time of year and had peaked on October 21 in 1973. Nonetheless, Coyne and his crew won the *National Enquirer* Blue Ribbon Panel's award for "the most scientifically valuable report of 1973." The award was $5,000.

1. THEY DIDN'T LOOK LIKE JETS

Just when we're getting scientific enough to appreciate the possibility of traveling from one place to another, here come the flying saucers. R.R. Feynman, *The Meaning of it All: Thoughts of a Citizen Scientist*

They can call me Einstein, Flash Gordon, or just a screwball, but I am absolutely certain of what I saw. Kenneth Arnold, responding to press ridicule

It is a capital mistake to theorize in advance of the facts. Sherlock Holmes

It all began on a splendid, sunny, summer afternoon. The date was June 24, 1947, and on that pleasant afternoon, without warning, a

9

phenomenon that would have worldwide repercussions was about to begin.

Kenneth Arnold, a thirty-two year old businessman from Boise, Idaho, was flying his single-engine plane from Chehalis to Yakima. Flying at 9,200 feet over the Cascade Mountains in Washington, he was admiring the awesome view. He had been installing fire-fighting equipment for the Chehalis Central Air Service that afternoon when he heard there was a $5,000 reward for anyone who could locate the wreckage of a US Marine Corps C-46 transport that had tragically gone down somewhere in the Mt. Rainier area.

The crash site was somewhat off his planned route, but he decided to make the short detour to see if he could spot the wreckage. As he cruised along, a sudden blue-white flash startled him. He immediately thought "Explosion!" Perhaps another aircraft also searching for the crash site had exploded near him. The flash seemed close. He glanced at his instrument panel. The clock showed almost three o'clock. He stiffened and waited for the sound and shock wave from the blast. But nothing happened. A few more seconds passed. Nothing. He scanned the sky in all directions. The only other airplane he could see was a DC-4 passenger plane that was far to his left and rear, apparently on its San Francisco-Seattle run.

Arnold began to think he must have been mistaken about the light when another bright blue-white flash lit up his cockpit. This time he realized the light had come out of the north in the vicinity of Mt. Baker. As he looked toward the north of Mt. Rainier, he saw something he had never seen before: nine odd looking objects skimming along the mountaintops at what looked like an incredible speed. They were flying in a tight echelon formation. Every few seconds, two or three of them would dip or bank or change course slightly. Whenever they performed these maneuvers, their mirror-like surfaces would reflect a bright flash of sunlight. The objects appeared to be crescent-shaped with no tails, and they flew with an odd weaving motion, unlike any aircraft Arnold had ever seen.

He figured their wingspan to be 45 to 50 feet, and he calculated their speed at an incredible 1,656 mph. Initially assuming they were some type of new aircraft, he now realized their speed was nearly three times faster than any jet. He would later say he never saw anything fly so fast. Yet he saw no vapor trails. It was hard to judge their distance, but he thought they were 20 to 25 miles away from him when he first saw them. He thought they were about the size of a DC-4 airliner. When seen edge-on they looked like a straight black line.

Arnold landed in Yakima around 4:00 PM. He told his friend Al Baxter, the manager of Central Aircraft, about the incident. Baxter called in several of his pilots to listen to Arnold's strange tale. One of the pilots thought the objects might be guided missiles from a nearby test range. But the objects had flown in formation; they had banked and turned. Missiles and rockets didn't do that. Then they must have been a new type of jet. But they didn't look like jets. They had no tails and they left no contrails. Their echelon formation was backward from that used by the air force. And they were flying much too fast to be jets.

Arnold took off for Pendleton, Oregon. He had tried to contact the FBI but the Yakima office had been closed, so he told the local media. The story had gone out over the wires of the Associated Press. When he arrived at Pendleton a swarm of reporters were waiting for him. Barraged with questions, some openly skeptical and even hostile, Arnold nevertheless stuck to his story: "I observed far to my left and to the north, a formation of nine very bright objects coming from the vicinity of Mt. Baker, flying very close to the mountain tops and traveling with tremendous speed. I could see no tails on them, and they flew like no aircraft I had ever seen before."

It was hard for the reporters to simply dismiss Arnold as a kook or hoaxer or publicity-seeker. The man was obviously a solid citizen. He was a successful salesman of fire-fighting equipment; he was an experienced search and rescue pilot, a deputy sheriff. He had logged more than 4,000 hours in the air, he flew his own plane, and he had flown over the Cascades many times. He had been an Eagle Scout. When pressed to describe the strange objects he had seen, Arnold tried to find the exact words. He said he thought they looked like speedboats in rough water, or perhaps like the tail of a Chinese kite blowing in the wind. Then he finally said, "They flew like a saucer would if you skipped it across the water." From this statement reporter Bill Begrette coined the term "Flying Saucers".

Some of the reporters were not satisfied with Arnold's

calculations. How did he estimate a speed of I ,656 mph? Such a speed was unheard off for any aircraft in 1947. Later when more precise calculations were made, the speed was estimated at no lower than 1,350 mph, which was still unbelievable. The objects were too fast to be jets but they didn't fly anything like a rocket or missile. So what were they?

Skeptics from the scientific community concluded he must have made an honest mistake. Perhaps he had misidentified a natural object or a conventional aircraft. He had probably misjudged the distance; the objects really had been much closer than he realized. Military jets flying at subsonic speed would have appeared incredibly fast at close range. Because he had used the mountains as a fixed reference point his estimate of size was wrong. The objects were actually much bigger than he had calculated. Most likely, they were bombers.

The Army Air Force (there was no USAF at this time) refused to say whether it did or did not have any aircraft in the vicinity on that day. An army spokesman commented, "As far as we know, nothing flies that fast except a V-2 rocket, which travels too fast to be seen." The official conclusion was the whole thing had been an optical illusion or mirage. The tips of the mountains had appeared to float above the earth as a result of a layer of warm air.

The press was merciless and for months Arnold was ridiculed as "the man who saw the Men from Mars." He claimed he would never report another sighting, "even if it were a 10-story building flying through the air." This comment by Kenneth Arnold, the first credible witness to report a UFO sighting, would echo eerily through the ensuing years and decades as other witnesses came to bitterly regret reporting their UFO sightings.

Within a few days of Arnold's sighting at least 20 other people came forward, claiming that they too had seen puzzling objects in the sky. Some of these sightings were reported on the same day as Arnold's, some had preceded it, and some came a day or two later. During the week of July 5 as many as 100 sightings a day were reported.

Although the authorities tended to dismiss these sightings, the fact that Ken Arnold was a respected businessman, deputy sheriff and pilot made it difficult to just brush off his report. Many people began to suspect some kind of cover-up. Apparently changing his mind, Arnold later collaborated with Ray Palmer, the editor of *Fate* magazine. In the premier issue of the magazine (Spring 1948) Arnold's first person account of his encounter with the nine "flying disks" was featured as the cover story. Arnold and Palmer later wrote a book in 1952 that included a more detailed account of the sighting.

Four days after Arnold's sighting a P-51 pilot flying near Lake Mead, Nevada reported a formation of five or six circular objects off his right wing. That very same evening, near Maxwell Air Force Base in Montgomery, Alabama, several air force officers saw a bright light zigzagging across the sky at high speed. The light flew overhead, made a 90 degree turn, and then disappeared to the south. A report came from White Sands Proving Ground in New Mexico of a pulsating light that flew from horizon to horizon in 30 seconds.

On July 2, only eight days after Arnold's sighting, perhaps the most famous and legendary of all UFO incidents occurred. On that day a "large glowing object flying at high speed" was sighted around 9:50 PM near Roswell, New Mexico. Mac Brazel, a local sheep rancher, said he heard a tremendous explosion that was much louder than the thunderstorm then sweeping through the area. In the morning Brazel found fragments of a silvery, foil-like material scattered over a quarter mile area. He described the material as thin but tough. Brazel also reportedly found a disk-shaped object that he turned over to the intelligence officer at Roswell Army Air Field.

This army airfield was home to the elite 509th Bomber Group. During World War II the B29 bombers that carried the atom bomb originated here. In 1947 Roswell was also home to the 33rd Fighter Group which was equipped with P-51 Mustangs. On July 8 the *Roswell Daily Record* quoted Lieutenant Warren Haught, a public relations officer at the base: "The many rumors regarding the flying disk became a reality yesterday when the intelligence office of the 509th Bomb Group of the Eighth Air Force, Roswell Army Air Field, was fortunate enough to gain possession of a disk through the cooperation of one of the local ranchers and the sheriff's office of Chaves County." The report caused a sensation that continues to this

13

day.

The publicity surrounding Arnold's sighting plus the continuing reports of strange and unusual objects in the skies that seemed to be flying at fantastic speeds fueled the public imagination. Already, more people knew about flying saucers than about the Marshall Plan that had been drafted to rebuild post-war Europe. Flying saucer enthusiasts, who had been accused of dreaming, hallucinating, imagining or worse, felt vindicated. At , here in Roswell, was physical proof that flying saucers were real.

The debris from the Roswell wreckage was loaded on an aircraft and taken to Carswell AFB near Fort Worth, Texas. There Brigadier General Roger Ramey announced on the radio that it had all been a mistake. According to Ramey, Mac Brazel had simply found the remains of a crashed weather balloon. The air force refused to elaborate further. A press conference was held where General Ramey repeated his assertion that the debris was from a weather balloon. He even allowed photographers to take some pictures of the wreckage. (Some of these photos later became classics).

But the reporters, not to mention the public, were not satisfied. They complained that they had not been allowed to get close enough to the debris. A second press conference was held. This time one of the photographers claimed the wreckage looked different. He accused the air force of switching the fragments.

Meanwhile back in Roswell, local residents who had seen or otherwise been in contact with the strange material said they were being ordered by federal agents to remain quiet about what they had seen and, in some cases, touched. Some later claimed that their very lives had been threatened if they talked to reporters.

All of these measures satisfied no one. The air force was accused of a massive cover-up of a crashed saucer that proved beyond doubt that Earth was being visited by aliens from another world. Forty years later documents would surface alleging that four dead aliens had been retrieved from the wreckage, and they and the remains of their craft were being held in a secret government facility. An autopsy was even supposed to have been performed on one of the dead crew beings.

In the midst of all the hoopla and wild accusations and rumors, credible sightings continued to be reported by credible people. On July 7, 1947 an observer near Phoenix, Arizona saw an elliptical, flat, gray object about 20 to 30 feet across and flying between 400 and 600 mph. It then spiraled downward from 5000 feet to 2000 feet. The witness ran into his garage, grabbed a camera, and snapped two pictures. The subsequent photos showed an object that appeared to be jet black with sharp outlines. A government study, called Project Grudge, later judged this report to be "too vague for interpretation." The same conclusion was later made about another sighting (Incident #51, Condon Report) made by a homemaker in Oswego, Oregon who saw 12 to 15 round silver objects flying at high altitudes on September 3, 1947.

Sightings continued into the following year. On January 7, 1948 at 1:15 PM the Kentucky State Highway Patrol was receiving reports from ordinary citizens about a large saucer-shaped object in the sky. Witnesses were saying it was about 200 to 300 feet in diameter. This object was also seen at the control tower of Godman AFB outside Louisville. A group of four F-51 Mustangs arrived at the same time. The base commander asked the flight leader, Captain Thomas Mantell, to investigate.

Captain Mantell and two of the other pilots set out. At first they couldn't see the object and they followed directions from the control tower. Captain Mantell then got a visual sighting and reported the object was ahead of him, but flying higher, so he was climbing to 20,000 feet. He radioed back, "I'm going to 20,000 feet and if I'm no closer then, I'll abandon the chase." The two other Mustangs stayed behind as none of the pilots had oxygen. Although they tried to radio Mantell again, he never responded. At 4:00 PM it was reported that his aircraft had crashed and he had been killed.

Captain Thomas Mantell became the first person to be killed in a UFO chase. Up to this time UFO sightings with possibly the exception of Roswell had .been regarded more or less as a harmless pastime. But Captain Mantell's death added a sinister, foreboding element to the phenomenon. Suspicions began to grow that UFOs might be more dangerous than the government was willing to admit.

It was later concluded that the mysterious object Captain

Mantell had been chasing was either the planet Venus or a "skyhook" balloon that the navy had been secretly flying as part of a high altitude research program. Yet it was hard to believe that an experienced air force pilot would misidentify Venus or a research balloon. Once again, the public suspected a cover-up. Rumors circulated that the UFO had deliberately shot down Captain Mantell.

On July 24, 1948 the two pilots of an Eastern Airlines DC-3 saw a bright light straight ahead of their aircraft and coming rapidly right toward them. They pulled their plane sharply to the left to avoid a collision. When they looked back, the UFO was going into a steep climb. The pilots later said it looked like a wingless B-29 with an underside that was glowing a deep blue.

On October I Lieutenant George F. Gorman of the North Dakota National Guard was flying a Mustang F-51. As he approached Fargo, North Dakota and prepared to land, he suddenly saw the taillight of another airplane dart by him on the right. But the control tower said there were no other planes in the area except for a Piper Cub that was below him. Despite what had happened to Captain Mantell, Gorman decided to chase the mysterious light.

"It was about six to eight inches in diameter, clear, white, and completely round, with a sort of fuzz at the edges. It was blinking on and off." As he approached the light, it suddenly veered away in a sharp left turn and dove. Gorman threw his Mustang into a 400-mph dive, but the object suddenly reversed its course and started a steep ascent. He continued after the UFO, but it made a sudden, sharp, right turn. The aircraft and the UFO were heading straight for each other and were about to collide when Gorman again threw his Mustang into a dive. He saw the object pass about 500 feet over him. Again he went after the UFO and once again it turned and flew straight at him. But just before they were about to collide, the object went into a vertical climb. Gorman tried to follow it, but at 14,000 feet his Mustang lost airspeed, shuddered, and stalled. The UFO headed north by northwest then shot out of sight. The encounter had lasted for 27 minutes. Gorman later said he felt the UFO was "controlled by thought. The maneuvers were just too sharp and too swift to have been performed otherwise."

According to the Condon Report the light had been blinking, but then suddenly became steady as Gorman got within 1,000 yards of

16

the object. The Air Weather Service later said Gorman had been chasing a lighted balloon.

The immediate post war years drew to a close. It was 1950 and the new decade began with a UFO report that would later become one of the classic sightings among ufologists. The incident became Case # 46 in the Condon Report.

On May II, 1950 in McMinnville, Oregon Mrs. Evelyn Trent saw a metallic-looking, disk-shaped UFO. She called to her husband Paul who got their camera and managed to photograph the object before it disappeared. It was about 7:30 in the evening and Mrs. Trent was outside in the backyard feeding her rabbits. The object was gliding slowly and silently overhead.

While Trent photographed, the couple watched the object as it hovered almost directly above them. It then shifted its position and orientation, changing direction and tipping, just before it moved away. It didn't seem to undulate or rotate, just glided. Shortly after Trent took the second photo, the object accelerated slowly and moved rapidly toward the west.

The Trents described the object as "very bright -- almost silvery, brightly metallic, silver or aluminum colored, with a touch of bronze ... it appeared to have a superstructure .. .like a good-sized parachute canopy without the strings, only silvery-bright mixed with bronze." The object made no noise, and there was no visible exhaust, flames, or smoke.

What made the McMinnville sighting so credible were the photographs. Although the Trents did not seek publicity, a McMinnville newspaper reporter got hold of the negatives and broke the story. On June 8, 1950, the two photos appeared on the front page along with an article and Editor's Note:

.. .In view of the variety of opinion and reports attendant to the saucers over the past two years, every effort has been made to check Trent's photos for authenticity. Expert photographers declared there has been no tampering with the negatives. The original photos were developed by a local firm. After careful consideration, there appears to be no possibility of hoax or hallucination connected with the pictures. Therefore the "Telephone Register" believes them authentic ...

17

The Portland, Oregon and Los Angeles newspapers also carried the story, and the following week *Life* magazine published the photos. The Trents went to New York to appear on the TV show *We the People*.

After investigating the incident, interviewing the Trents and other McMinnville residents who knew them, and examining the original negatives, the Condon Report concluded:

This is one of the few UFO reports in which all factors investigated, geometric, psychological, and physical appear to be consistent with the assertion that an extraordinary flying object, silvery, metallic, disk-shaped, tens of meters in diameter, and evidently artificial, flew within sight of two witnesses. It cannot be said that the evidence positively rules out a fabrication, although there are some physical factors such as the accuracy of certain photometric measures of the original negatives which argue against a fabrication.

McMinnville was only the beginning. The decade of the 1950s saw such a rash of UFO reports that the phenomenon turned into a national craze. In fact, the 1950s were probably the height of the flying saucer mania. Ray Palmer's *Fate* magazine became a leader in publicizing UFOs. But there were plenty of other magazines and books that jumped on the bandwagon. Hollywood capitalized on the craze by producing a plethora of cheesy flying saucer movies with occasional well-made, thoughtful ones like *The Day the Earth Stood Still* (195 I) and *War of the Worlds* (1953). It seemed that the majority of Americans were convinced the flying saucers were not only real but came from other worlds. The only debate was whether the space people's intentions were good or evil. Were they a threat to us? Should we fear them? Or had they come to help us? Perhaps they came to rescue us from the threat of nuclear annihilation?

For some people UFOs took on an almost mystical, religious significance. Believers began to hold conventions, particularly in California at gatherings like the Giant Rock Convention. People told of making contact with alien beings. Some, like George Adamski, told incredible tales of being taken aboard flying saucers and going for rides through the solar system. Adamski's contacts were Firkon of Mars, Ramu of Saturn, and Orthon of Venus.

Skeptics who scoffed at the whole flying saucer fad claimed that most UFO sightings were nothing more than honest misidentifications of natural phenomenon such as bright planets, ball lightning, geese flying in formation and reflecting light, or else misidentifications of known objects like the ever reliable weather balloon or aircraft that happened to be seen in unusual atmospheric conditions. The skeptics also pointed to reports that were obvious hoaxes, although they tended to be charitable toward these cases. While acknowledging that some UFO reports did seem to be genuine unknowns, they took the position that there was not enough data to draw a definite conclusion.

On September 10,1951 an ANIMPG-1 radar set picked up a high speed, low-flying object southeast of Fort Monmouth, New Jersey. The target appeared to be following the coastline. While it changed its range only slightly, its azimuth changed rapidly. The radar set was switched to full-sided azimuth tracking, usually fast enough to track jet aircraft, but not in this case. Full-sided azimuth tracking could track any aircraft flying at speeds of up to 700 mph, but since the radar operator could not track the object, he assumed its speed was in excess of 700 mph.

The next day there was another sighting at Fort Monmouth. This time a target was picked up on an SCR-584 mobile radar set. At first, the target remained stationary on the scope and appeared to be hovering. The operators looked out of their van to see if they could see the object visually, but the sky was too overcast. Returning to their operating positions, they watched the object suddenly and very quickly change its elevation, appearing to rise nearly vertically. The speed of the target was too fast for the radar set and once again, the operators concluded, the object was faster than 700 mph. Project Blue Book would later dismiss the object as "an example of anomalous propagation." Also, the radar operators were aware of the radar sighting from the previous day and were therefore, "in the correct psychological condition to see more such objects. "

By the beginning of 1952 public interest in UFOs had

reached the point where many people believed they would never get the truth from their government, and so they formed their own amateur study groups to try to solve the mystery. The first of these was the Aerial Phenomena Research Organization (APRO) of Tucson, Arizona founded by James and Coral Lorenzen. In 1966 Coral Lorenzen published *Flying Saucers: The Startling Evidence of the Invasion from Outer Space*. The central premise of her book was that the government establishment was covering-up its knowledge of flying saucers. By 1968 APRO claimed 8,000 members.

During 1952 a whopping 1,501 reports of flying saucers came from all across the United States. The media of the day continued to fuel the movement by publishing hundreds of popular articles. And it wasn't just *Fate* magazine anymore. Magazines that published UFO articles that year included *American Mercury, Collier's, Life, New Republic, Newsweek, New Yorker, Popular Science, Reader's Digest,* and *Time.* Clearly, UFOs had entered the American mainstream. But the biggest UFO event of 1952 was the Great Washington D.C. Saucer Flap.

What would become one of the most credible and genuinely mystifying of all UFO sightings first occurred on July 19, 1952. Air Controllers at Washington's National Airport picked up seven slow-moving objects on two radarscopes. Radar showed the objects to be about 15 miles from the airport and flying between 100 and 130 mph. Fifteen miles away, controllers at Andrews AFB in Maryland were seeing the same blips on their radar.

According to the Condon Report, one of the airmen at Andrews reported the following:

"Airman X called the tower and reported he had seen objects in the air around Andrews; while we were discussing them he advised me to look to the south immediately. When I looked, there was an object which appeared to be like an orange ball of fire, trailing a tail; it appeared to be about two miles south and one half-mile east of the Andrews Range station. It was very bright and definite, and unlike anything I had ever seen before. The position of something like that is hard to determine accurately. It made kind of a circular movement, and then took off at an unbelievable speed; it disappeared in a split second. This took place around 0005 EST.

Seconds later, I saw another one, same description as the one before; it made an arc-like pattern and then disappeared. I only saw each object for about a second. "

This is the account of a staff sergeant at Andrews AFB: "Later on, we spotted what seemed to be a star northeast of the field, which was in the general direction of Baltimore. It was about tree top level from where I was watching. It was very bright, but not the same color as some apparent meteors. This was a bluish silver. It was very erratic in motion; it moved up from side to side. Its motion was very fast. Three times I saw a red object leave the silver object at a high rate of speed and move east out of sight. "

At 3:00 AM Harry Barnes, the senior controller at National, officially notified the US Air Force Air Defense Command. One half hour later a pair of radar equipped F-94 night fighters roared in and the blips disappeared from the scopes. The jets left. As soon as they were gone, the blips reappeared as if by magic.

On July 26 the UFOs returned. At 10:30 PM air traffic radar at National Airport once again picked up the strange blips. Once more Harry Barnes checked with Andrews AFB and once again the controllers there confirmed the unknown targets were also on their scopes. The blips showed five or six objects that seemed to be moving south. Meanwhile, commercial airline pilots were also radioing in reports of strange objects sighted near the airport.

At 11:00 PM Barnes called the Pentagon and at II :25 a pair of F-94s were in the air. As before, the UFOs instantly disappeared from the radar screens. As soon as the jets were gone, the UFOs were back. At 3:20 AM the air force sent in another pair of F-94s. This time the UFOs didn't disappear, and one of the jet fighter pilots reported a visual sighting of four lights. At one point the pilot radioed that the lights were surrounding his plane. Before the ground controllers could respond, the lights disappeared.

The next morning even President Harry Truman was demanding an explanation. Finally, on July 29 Major General John A. Samford, director of air force intelligence, held a press conference and told reporters that all the sightings over the Washington D.C. area during the past two weeks were caused by temperature inversions.

Despite such blase public assurances, the U.S. Air Force as well as other branches of the military and the government were privately beginning to worry. People were obviously seeing strange objects in the skies. It was bad enough when ordinary citizens reported UFOs, but it was especially worrisome when the sightings came from radar operators, commercial airline pilots and military pilots. Trying to explain these sightings as bright planets, research or weather balloons, mirages, and temperature inversions must have sounded ludicrous even to their ears. They knew the public wasn't buying it and they feared a panic. They worried that UFO sightings might threaten the stability of the government and perhaps the security of the nation. What if communication lines were jammed with UFO reports at the very moment the Soviet Union was launching a secret attack? The American public must be reassured; they must be convinced that UFOs were neither a threat to themselves or to the nation's security.

But they couldn't be told the truth

2. FROM SIGN TO THE CIA

Where secrecy is known to exist, one can never be absolutely sure that he knows the complete tmth.　　And *I'm almost inclined to think such studies ought to be discontinued unless someone comes up with a new idea on how to approach the problem ... The 21st century may die laughing when it looks back on many things* we *have done. This* (his UFO study) *may be one.*
--Dr. Edward U. Condon

The United States Army's response to the first UFO sightings was to examine flying saucer reports at the Air Technical Intelligence Center (A TIC) at Wright-Patterson AFB near Dayton, Ohio. On September 23, 1947, Lieutenant General Nathan F. Twining, Army Chief of Staff, sent a memo to the Commanding General of the Army Air Force (there was no USAF yet) directing him to establish an official study of flying saucers. On December 30, 1947 the study was given the code name Project Sign. Two years later it would be released to the public under the name Project "Saucer."

Even at this early date, many of the attitudes Americans have today toward UFOs were already apparent. Some people were convinced that flying saucers were interplanetary craft piloted by beings from other worlds, others thought they might be Soviet secret weapons, skeptics thought they were misidentifications, hallucinations, or hoaxes. As in all subsequent studies, Project Sign was quick to reassure an anxious public that "exhaustive investigations have turned up no alarming probabilities" and that the idea of flying saucers being from outer space was "looked upon as an almost complete impossibility."

Army Air Force attitudes were somewhat different. Some thought the whole flying saucer craze was ridiculous and wanted nothing to do with it. But other military officials were concerned that the unidentified flying objects might belong to a foreign power (read the Soviet Union) that was planning an invasion of the U.S. Perhaps flying saucers were a prelude to the main attack. Those who thought along these lines also did not entirely rule out the possibility of an alien invasion from outer space. No doubt, there was concern among these military minds that many of the flying saucer reports came from pilots, considered to be the most reliable and credible of witnesses. If these experienced military and commercial pilots were convinced they were seeing flying objects that were extremely unusual, there just might be something to it. UFOs shouldn't be merely brushed aside or ignored.

Lieutenant General Twining wrote in his memo:

It is the opinion (of this command) that the so-called flying saucers phenomenon is something real and not visionary or fictitious ... The reported operating characteristics such as extreme rates of climb, maneuverability and evasive action when sighted or contacted by friendly aircraft and radar lend belief to the possibility that some of the objects are controlled either manually, automatically or remotely."

It is important to note that Project Sign began in 1947, only two years after the end of World War II which saw the emergence of highly developed machines and weapons of war. The war, which. began in Europe in 1939, ended six years later with V-2 rockets, the first jets, the first guided missiles, and the explosion of two atomic bombs. During the war the U.S. and the Soviet Union had been allies, but the relationship quickly fell apart after the defeat of Nazi Germany. The Soviet Red Army that had liberated Eastern Europe from Nazi domination stayed in place after the war. The ruthless Soviet Dictator, Joseph Stalin, began moves to eradicate all opposition to communism and to consolidate the Soviet position throughout the region. On March 5, 1946 Great Britain's wartime Prime Minister, Winston Churchill said in a speech in Fulton, Missouri that an "iron curtain" had descended across Europe.

In 1949 the U.S. and her allies established the North Atlantic Treaty Organization (NATO) to counteract the growing communist threat. In response, the Soviet Union and her reluctant allies formed the Warsaw Pact. As the two blocs eyed each other suspiciously and nervously they soon created the long, hostile, and very dangerous stalemate that became the Cold War.

In 1947 President Harry Truman outlined his Truman Doctrine. Among other things it "supported free peoples who are resisting attempted subjugation by armed minorities or outside pressures." On March 12 Congress granted $400 million to aid Greece and Turkey in their fight against communist guerrillas. The National Security Act was established, creating the Department of Defense and several new agencies including the National Military Establishment with three separate departments of the Army, the Navy, and the newly created United States Air Force. In addition, the National Security Council, the Central Intelligence Agency (CIA) and the Joint Chiefs of Staff were formed. Of these, the USAF and especially the CIA would become the most involved in the UFO issue.

Project Sign investigated many of the early flying saucer sightings, particularly those reported by radar operators and pilots. In February 1949 the name of the project was officially changed to Project Grudge which indicated the air force's attitude toward the UFO phenomenon.

Project Sign's final report while citing the necessity for getting more factual evidence on sightings and more definite data concluded, "these sightings do not represent a threat to the security of the nation." Project "Saucer" did admit, "there are still question marks in the

Saucer Story." Military pilots who saw anything anomalous were advised to notify neighboring bases so that other observers, in flight or on the ground, could help with identification.

Thus, the first official report on UFOs set the tone for all the projects and studies to come. Flying saucers were not a threat to national security, but evidence and data on them should still be gathered and analyzed. However, the staff of Project Sign supposedly prepared an "Estimate of the Situation" which was classified top secret. The Estimate claimed the staff was convinced that the flying saucers were interplanetary space vehicles. General Hoyt Vandenberg, the Air Force Chief of Staff, reportedly refused to accept such a conclusion on the grounds that the Estimate lacked sufficient proof to draw this conclusion. As a result, the Estimate never became an official document of the air force. Copies of the Estimate were destroyed, although a few copies were thought to have been hidden away. The University of Colorado Project was unable to later verify the existence of the Estimate. Nonetheless, it made the same conclusion as General Vandenberg: the evidence was "inadequate to support such a conclusion."

When Project Sign began the expertise of J. Allen Hynek was enlisted. Dr. Hynek was a professor of astronomy at Ohio State University, and was chosen as the scientific advisor to help evaluate the reports coming in. He was highly skeptical about UFOs and later remarked that the air force seemed as impressed by his skepticism as by his academic credentials. Hynek would remain as the scientific consultant for the subsequent Project Grudge and Project Blue Book, although he later claimed he played no part in Grudge.

Twice a month Professor Hynek would drive from Ohio State to Wright field to read through a stack of UFO reports. He later said he always started with the assumption that there was a natural explanation. In addition to natural phenomenon, like the planet Venus or ball lightning, he attributed some sightings to misidentifications of military aircraft or to research balloons. These balloons were used extensively throughout the 1940s and 1950s. They were made of a translucent plastic that could glisten or change colors in sunlight. Some were top secret and were used as spy balloons for floating over the Soviet Union. In addition, the military was using some odd-looking aircraft at the time. One test plane developed by the navy was wingless with a saucer shape. Circular wing aircraft were also being developed and tested, the theory being a circular shaped craft could best avoid radar detection.

In 1966 Hynek became embroiled in a UFO flap that occurred near Ann Arbor, Michigan. On March 14, police officers and ordinary citizens in three Michigan counties said they saw lighted objects flashing across the early morning skies. One deputy sheriff reported: "these objects could move at fantastic speeds, make very sharp turns, dive and climb and hover with great maneuverability."

Three days later, other witnesses reported the same thing. On March 20, Frank Mannor, a forty-seven year old truck driver said: "We got to about 500 yards of the thing. It was sort of shaped like a pyramid, with a blue-green light on the right hand side and on the left a white light... The body was like a yellowish coral rock and looked like it had holes in it ... You couldn't see it too good, because it was surrounded with heat waves, like you see on the desert. The white light turned to a blood red as we got close to it ... " The next day, the strange lights were seen in the sky again, this time by more than 140 witnesses."

Hynek, now a consultant to Project Blue Book, was sent to Ann Arbor to investigate. He complained that the "situation was so charged with emotion that it was impossible for me to do any really serious investigation." He also complained that "clusters of reporters" were obstructing his work and that he was getting no help at all from the air force. He concluded his Ann Arbor investigation by speculating that the UFOs might be swamp gas. This touched off a furor and gained notoriety as the "swamp gas affair." Hynek later called the whole episode "the low point of my association with UFOs."

Although Hynek began his UFO career as a confirmed skeptic, his attitude began to change during the 1960s. He felt that "The cumulative weight of continued reports from groups of people around the world whose competence and sanity I have no reason to doubt" pointed to the possibility of a different . He said, "I have not always held the opinion that UFOs were worthy of serious study. I began my work as Scientific Consultant to the U.S. Air Force as an open skeptic, in the firm belief that we were dealing with a mental aberration and a public nuisance. Only in the face of stubborn facts ... have] been forced to change my opinion."

On February II, 1949 Project Sign officially became the aptly named Project Grudge. Grudge studied 244 sighting reports. About 32 percent of them were explained as misidentifications of natural objects, 12 percent were thought to be sightings of those good old weather balloons, 33 percent were dismissed either as hoaxes or reports that were too vague for any kind of explanation, and the remaining 23 percent were classified as genuine unknowns. The 1969 University of Colorado Project that received a copy of Project Grudge thought that many of the sighting reports were too vague for interpretation and that the air force investigators had actually given them more credence than they deserved.

Among other recommendations, Project Grudge urged the "study of reports of unidentified flying objects be reduced in scope." In conclusion the project stated:

There is no evidence that objects reported upon are the result of an advanced scientific foreign development; and therefore they constitute no direct threat to the national security. In view of this, it is recommended that the investigation and study of reports of unidentified flying objects be reduced in scope. Headquarters AMC will continue to investigate reports in which realistic technical applications are clearly indicated ... All evidence and analysis indicate that reports of unidentified flying objects are the result of 1. Misinterpretation of various conventional objects 2. A mild form of mass hysteria and war nerves 3. Individuals who fabricate such reports to perpetrate a hoax or to seek publicity and 4. Psychopathological persons.

In December 1948 the report was put in storage and the project's personnel were transferred to other jobs. Apparently, Project Grudge had been designed to put the kibosh on any more UFO studies but not before it made two critical mistakes that were a harbinger of things to come.

The projects conclusions, either intentionally or unintentionally, tended to label witnesses as unreliable at best and

27

crazy or pathological at worse. People who reported seeing strangeness in the skies ran the risk of being called "strange" themselves. Too often skeptical reaction was to question the background and motives of witnesses, even their psychological makeup, instead of taking a cold, hard, objective look at the objects the witnesses were reporting. Shifting attention from the UFOs to the witnesses who saw them was perhaps the biggest and arguably the most tragic mistake made by the investigators, and it set an unfortunate precedent that was to forever plague UFO research. Not only were reputations tarnished and sometimes even ruined, but this stubborn and inexplicable attitude toward the witnesses prevented investigators from moving forward and making progress. Critics who would later complain about the "static" nature of UFO research failed to make this essential connection. Grudge's other major error was to classify reports as secret, especially the "good ones," thus preventing independent scientists from conducting their own studies and comparing notes. Also, classifying reports as secret roused suspicions, and gave rise to all sorts of conspiracy theories.

In the early days of flying saucer sightings from around 1949 to 1952 many people involved in UFO investigations came to believe the extraterrestrial hypothesis was the only theory that seemed to fit the facts, especially because a lot of pilots, both commercial and military, were reporting sightings of UFOs. Not only were their reports highly credible, they were remarkably consistent and similar: the UFOs were described as "material objects," they were "metallic," they flew at "fantastic speeds," they appeared to take "evasive action," and there were enough reports of UFOs being tracked on radar to refute the notion they were hallucinations or mirages.

In March 1952 Project Grudge was changed to Project Blue Book. Although Grudge had recommended reducing and minimizing UFO investigations, the air force was well aware it still had a mammoth public relations problem. So for the next seventeen years, from 1952 to 1969, the air force would ostensibly study UFOs. To show their sincerity, Captain Edward J. Ruppelt was appointed director. He had been a bombardier during World War II, and was known for his level head. He was determined to be objective. Professor Hynek was retained as the scientific consultant. In 1956

Captain Ruppelt published *The Report on Unidentified Flying Objects.*

After Captain Ruppelt, Lieutenant Colonel Hector Quintanilla headed Project Blue Book. The project would investigate a total of 12,618 sightings with 701 of these remaining unknown. Like Sign and Grudge before it, Project Blue Book concluded that UFOs presented no threat to our national security. On December 17, 1969 the Department of Defense announced that the air force was disbanding Project Blue Book. Officially the air force was no longer interested in UFOs, and the U.S. government was washing its hands of the whole business.

Although he was Blue Book's scientific consultant, J. Allen Hynek was highly critical of the project. He claimed the Blue Book Theorem was "It can't be, therefore it isn't." It was" the Society for the Explanation of the Uninvestigated." Hynek charged that Blue Book never did sufficient follow-up, and no case was given "the FBI treatment" where every bit of evidence is gathered and analyzed. Cases that were likely misperceptions or hoaxes were intensively investigated while truly puzzling cases were virtually ignored.

Oftentimes, cases were not investigated until they were reported in the popular press. Only a Congressional inquiry could galvanize the investigators out of their torpor. "Nothing brought more immediate and frenzied reaction from Blue Book than a query from Congress," said Hynek. Then a quick, but satisfactory answer was given in lieu of a serious study. There was no scientific dialog between Blue Book and the outside scientific world. The staff, both in numbers and scientific training, was grossly inadequate. Statistical methods, data gathering, methodology, and interrogation of witnesses were poor. Hynek was especially upset that UFO reports from other countries were routinely dismissed even when they dovetailed with U.S. sightings.

In 1968 after 20 years as a scientific consultant to the air force, Hynek was officially asked for criticism and advice. In response, he outlined many of the weaknesses and problems he had encountered over the years. But instead of taking his advice and considering his recommendations, the air force shut down Blue

Book one year later. Hynek concluded that either the Pentagon knew UFOs really didn't pose a threat to national security or if they did, the problem was being taken care of elsewhere.

After Project Grudge grudgingly became Project Blue Book the CIA became interested in UFOs. The newly created agency (in 1952 the CIA was only five years old) feared that military communication channels might be jammed with UFO sighting reports at the very moment an enemy was launching a sneak attack on the U.S. Also an enemy prior to launching such an attack might deliberately generate a wave of UFO reports in order to jam these channels. However, it wasn't long before the CIA realized it could use the UFO phenomenon to its own advantage.

By 1953 the Cold War between the United States and the Soviet Union was entrenched. All of the countries of Eastern Europe were now controlled by Communist regimes. In 1949 The People's Republic of China had been established, NATO was formed, and the Soviet Union exploded an atomic bomb. In 1950 Senator Joseph McCarthy began his Communist witch-hunt, and the Korean War began. In 1953 the Soviet Union exploded a hydrogen bomb. Thermonuclear war between the two superpowers seemed likely to break out at any moment. One of the CIA's main concerns was how to monitor the Soviet Union's nuclear activities, particularly its weapons testing. The U.S. wanted to "see" inside the tightly closed society that was the Soviet Union. This had direct implications for UFOs.

In March 1954 a huge flying saucer festival, the Giant Rock Convention, was held in California. Some 5,000 flying saucer enthusiasts attended. The convention was given wide coverage by the media of the day, including television. Some of the attendees claimed to have made hundreds of trips in spaceships with benevolent aliens who had come to save Earth from nuclear annihilation. They gave detailed descriptions of the spaceships, and some even made drawings of the aliens. One alien looked remarkably like *Star Trek's* Captain Kirk although this was years before the TV series.

After the Washington D.C. Saucer Flap in 1952, the CIA

created a panel of five scientists chaired by Professor H.P. Robertson of the California Institute of Technology. All of the panel members had backgrounds in physics research and had also contributed to military research during and after World War II. The panel spent one week in January 1953 studying selected UFO cases as well as viewing some of the UFO photos and motion pictures that were then available. However, it was not revealed to the air force or the public that the panel was under CIA sponsorship.

The Robertson Panel also reached the conclusion that "evidence presented on Unidentified Flying Objects shows no indication that the phenomena constitutes a direct physical threat to national security ... and that there is no evidence that the phenomena indicated a need for the revision of current scientific concepts." The panel recommended "that the national security agencies take immediate steps to strip the UFOs of the special status they have been given and the aura of mystery they have unfortunately acquired." Further recommendations included reassuring the public of the "total lack of evidence" behind the phenomena.

Not surprisingly, the panel came under immediate fire for only spending one week on a mystery that had been building for years. Allen Hynek complained that the panel was not given access to many of the genuinely puzzling cases, and it did not have enough material to work with. UFO enthusiasts were upset with its recommendation urging that UFOs be stripped of their "special status" and their "aura of mystery" while the panel's findings were classified "secret." (The findings were only partially declassified in 1966). Such secrecy and hypocrisy did nothing to boost public confidence in the government agencies that were supposed to be studying UFOs. This schizoid attitude of continuously dismissing UFO reports as nothing extraordinary, while maintaining an air of secrecy about them, cemented public belief in a government conspiracy. Even people who were not particularly into UFOs believed the government was covering-up and hiding Something.

In 1956 Donald Keyhoe, a retired Marine Corps major, founded the National Investigative Committee for Aerial Phenomena (NICAP). This amateur organization had two purposes: to study UFO cases and to attack the government for its handling of the whole UFO matter. Keyhoe had already written *Flying Saucers are Real* and now he published *Flying Saucers From Outer Space.* Another book during this period was *Flying Saucers on the Attack* by Harold T. Wilkins that posed the question: "Has the invasion of Earth by beings from another world already begun?" Wilkins also believed the government was hiding "astounding and incredible" facts from the public.

Clearly, the time had come for a full-fledged scientific study of UFOs. It finally came from the University of Colorado in the person of Dr. Edward U. Condon, a professor of physics and Fellow of the Joint Institute for Laboratory Astrophysics. From 1966 to 1968 the thirty-six member Condon Committee conducted "The Report the World Has Been Waiting For." But their conclusion offered nothing new, and was a major disappointment to UFO buffs. It was also heavily criticized by Dr. J. Allen Hynek, Dr. James E. McDonald who was the Professor of Atmospheric Sciences at the University of Arizona, and by Dr. David Saunders who was the project's Principal Investigator.

The project undertook the rather Herculean task of reviewing, exploring, and analyzing much of the UFO data that had accumulated over a twenty-year period. It looked at Projects Sign, Grudge, and Blue Book as well as the Robertson Panel. It analyzed a total of fifty-nine case studies. Cases I - 10 were studies prior to the term of the project, Cases 11 - 45 were studies during the term of the project, and Cases 46 - 59 were photographic case studies. The first case, number 46, was the famous McMinnville, Oregon photos.

The committee busily assembled a library, established investigative teams, and devised a method to study UFO reports. Although the project was supposed to be impartial and open-minded, Condon announced only three months after his appointment: "It is my inclination right now to recommend that the government get out of this business. My attitude right now is that there's nothing to it, but I'm not supposed to reach a conclusion for another year." Condon's chief administrator, Robert Low, held much the same attitude. In a now infamous memo to university officials Low wrote:

... Our study would be conducted almost exclusively by

nonbelievers who, although they couldn't possibly 'prove' a negative result, could and probably would add an impressive body of evidence that there is no reality to the observations. The trick would be, I think, to describe the project so that, to the public, it would appear a totally objective study, but, to the scientific community, would present the image of a group of nonbelievers trying their best to be objective, but having an almost zero expectation of finding a saucer. One way to do this would be to stress investigation, not of the physical phenomena, but rather of the people who do the observing -- the psychology and sociology of persons and groups who report seeing UFOs. If the emphasis were put here, rather than on examination of the old question of the physical reality of the saucer, I think the scientific community would quickly get the message ...

Before it was all over Condon would fire Dr. David Saunders and Dr. Norman Levine, two of the projects most active members, and Mary Louise Armstrong, Condon's Administrative Assistant, would resign.

Nonetheless, the University of Colorado Project is quite a read. The Bantam Book version published in 1969 runs over 900 pages and includes 24 appendices. The entire text of the Condon Report is also available on the Internet. The report contains a great deal of material about various UFO sightings, especially the early reports. After much "analysis" several explanations for UFOs were offered. A sampling: lenticular clouds, noctilucent clouds, mirages, sundogs, St. Elmo's fire, ball lightning, meteors, fireballs, satellite reentries, "skyhook" balloons, rocket launches, aircraft seen at unusual angles, aircraft landing lights, detergent foam, airborne debris, birds and flocks of birds, swarms of insects, and fireflies. There were also psychological/biological explanations: autokinesis, autostasis, entroptic effects, motes on the cornea, hallucination, excitedness effect, and airship effect.

Despite the excessive detail in the report, the project offered no new insights, interpretations, or explanations of UFOs. Its conclusions were consistent with those of Projects Sign, Grudge, and Blue Book and the Robertson Panel. On the very first page Condon wrote, " ... Our general conclusion is that nothing has come from the study of UFOs in the past twenty-one years that has added to our

33

scientific knowledge. Careful consideration of the record, as it is available to us, leads us to conclude that further extensive study of UFOs probably cannot be justified in the expectation that science will be advanced thereby."

Condon was mightily criticized for only devoting half his time to the project and then concentrating on the most bizarre and crackpot aspects of the UFO phenomenon. In his book *UFOs? Yes!* Saunders recounts some of the "nuts" Condon wasted his valuable time on a man who claimed he was in telepathic contact with extraterrestrials who were going to make a landing on Utah's Bonneville Salt Flats, an agent from the Third Universe who was also an extraterrestrial, and a man who claimed his wife's grandfather came from Andromeda. While Condon got a lot of laughs from the nut cases, Saunders felt it was neither professional nor healthy for the project, and it also offended many serious UFO investigators like Hynek, McDonald and Keyhoe. Commented Dr. McDonald; " The crackpots are so immediately recognizable that one need not waste any time at all on them."

Most disappointing of all, the Condon Report offered nothing new or different. J. Allen Hynek wrote in a review: " ... The book leaves the same strange, inexplicable residue of unknowns which has plagued the U.S. Air Force investigation for twenty years." The Report The World Has Been Waiting For was a bust.

What is so unbelievable about the University of Colorado Project is that nowhere in this voluminous report is the theory of top-secret aircraft ever even considered. It was as if the committee members had blinders on and refused to see what was right in front of them. Many of their case studies were obviously sightings of highly unusual aircraft, yet these were put down to lenticular clouds, aircraft landing lights, swarms of insects, detergent foam, motes on the cornea, hallucinations, and the like.

On January II, 1966 four observers in Meyerstown, PA witnessed a UFO for 10 minutes: "It was just like looking up under an airplane, just as if an airplane were standing there. Just perfectly motionless and no noise whatever. We watched this possibly for five minutes -- then the thing got a tremendous burst of speed and sped right off. No sound whatsoever." On June 8, 1966 in Kansas, Ohio there was a similar sighting: "There was no sound, and it was as long as a commercial airliner, but had no markings." And on May 24, 1965 a Trans-Australia Airlines pilot related: "I had always scoffed at these reports, but I saw it. We all saw it. It was under intelligent control, and it was certainly no known aircraft." How could the Condon Committee dismiss such reports?

Over the years, conspiracy buffs have suggested that Condon must have known about the secret aircraft and kept quiet about it, perhaps even deliberately misleading and obstructing his own panel. Yet, if you read his actual words, this doesn't seem to be the case. Although he possessed impressive scientific credentials, Dr. Edward Condon seemed to be inexplicably naive when it came to government secrecy and cover-up despite the fact that he had served under Robert Oppenheimer as Deputy Director of one phase of the Manhattan Project which produced the world's first atomic bomb. Condon was no stranger to top secret government projects.

Even Saunders refused to consider the possibility of secret aircraft. In *UFOs? Yes!* he wrote (and this is the one and only time he touched upon the subject): *We may as well deal right here and now with the possibility that flying saucers are man-made experimental aircraft or other devices ... The "Secret Project Hypothesis" can be meaningful only if it is worded something like, "All of the credible reports of truly remarkable UFO performance are due to classified military projects ... However, even with these qualifications, the hypothesis almost surely false. Any earthly government that could do in 1947 what flying saucers were reportedly doing then, and in public view, would have found it impractical to keep it secret for another 20 years. The economic pressures would be too great, and the military temptations would be too irresistible.*

Conspiracy buffs would later suggest that CIA operatives had infiltrated the Colorado project and deliberately steered the panel away from top secret aircraft. Project Coordinator Robert Low would come under heavy suspicion. At the time the Colorado Project was investigating UFOs the stealth program was underway, a project even more top secret than the Manhattan Project. According to one conspiracy theory, some of the kooks Condon was so taken with were in reality CIA agents who were deliberately discrediting the whole idea of UFOs by turning them into objects of ridicule and contempt, unworthy of serious, scientific study. But whether agents or crackpots or whether he knew of secret aircraft or did not, Condon by all accounts refused to take UFOs seriously and thereby undermined his own project.

After the Washington D.C. Saucer flap, the CIA formed a special study group within the Office of Scientific Intelligence (OSI) and the Office of Current Intelligence (OCI). Edward Tauss, acting chief of OSI's Weapons and Equipment Division, reported that most UFO sightings were nothing special and could be easily explained. Yet he recommended that the CIA continue to monitor the problem while concealing its interest from the media and the public. Such concealment only further contributed to the belief in conspiracies and cover-ups.

The CIA study group and air force officials met at Wright-Patterson AFB to review their data and findings. The air force believed that 90 percent of the sightings were nothing unusual, but the other 10 percent were believed to be credible ("incredible reports from credible observers" as Hynek put it). The air force's position sounded familiar: UFO reports were misidentifications of known objects or misinterpretations of natural phenomenon. The idea that they were secret weapons or "men from Mars" was rejected.

But H. Marshall Chadwell, Assistant Director of OSI, did not take UFOs so lightly. He wanted action to be taken because he was convinced "something was going on that must have immediate attention," the "sightings of unexplained objects at great altitudes and traveling at high speeds in the vicinity of major US defense installations are of such nature that they are not attributable to natural phenomena or known types of aerial vehicles." Chadwell learned that the British were also actively studying UFOs. Despite the fact that their conclusions were the same as the Americans, namely that UFOs posed no threat to national security, most Britons like their American counterparts were convinced that UFOs were real.

In January 1953 Chadwell helped put the Robertson Panel together. After the panel's conclusions, the CIA abandoned further

UFO studies although it continued to monitor sighting reports in the name of national security. But Chadwell was still not convinced. He was apparently worried that UFOs were some kind of secret Soviet weapon. The CIA knew that Project Y was a U.S., British, and Canadian effort to develop a non-conventional saucer type aircraft. Perhaps the Soviets were doing it too, and what if they produced theirs first?

In November 1954 the CIA began a top secret reconnaissance project for spy missions over the Soviet Union. The result was a high-altitude experimental aircraft known as the U-2. This aircraft could fly at 60,000 feet at a time when most commercial airplanes flew between 10,000 and 20,000 feet The first U-2s were silver and they reflected sunlight. (They were later painted black.) To observers below they could appear as fiery objects. The pilots who flew these planes resigned from the air force to work for the CIA. Project Blue Book investigators were aware of the secret spy plane and they knew that many UFO reports were actually sightings of the U-2. But they couldn't reveal this knowledge and so they said the UFOs were ice crystals or temperature inversions or weather balloons. Or they simply ignored some reports.

The U 2 Spy Plane

In addition to the U-2, the CIA was involved in other projects that involved highly classified aircraft. However, the very probable possibility that UFOs were top secret aircraft seemed to elude everyone who was ostensibly studying the UFO phenomenon.

On March 8, 1958 Donald Keyhoe, in an interview with Mike Wallace of CBS, claimed the CIA was deeply involved with UFOs and had actually sponsored the Robertson Panel. In the early 1960s Keyhoe and other UFO enthusiasts continued to pressure the agency to release its UFO information. They even called for congressional hearings and the release of all UFO materials. But the official line remained the same: UFOs were not a threat to national security nor were they of "foreign origin."

As public dissatisfaction mounted against Project Blue Book, the air force established a special ad hoc committee to review the project. Dr. Brian O'Brien, a member of the Air Force Scientific Advisory Board, and included Dr. Carl Sagan, the renowned astronomer from Cornell University, chaired the panel. Like all projects before it, the report concluded that UFOs were no threat to national security and it could find "no UFO case that represented technological or scientific advances outside of a terrestrial framework." But UFOs should still be monitored and studied.

In 1966 the House Armed Services Committee held hearings on UFOs. The results were the same. Harold Brown, the Secretary of the Air Force told the committee that most sightings were easily explained and there was no evidence that "strangers from outer space" were visiting Earth. But the air force would continue to investigate reports.

On a *CBS Reports* program Dr. Robertson revealed that the CIA had been involved in his panel's UFO studies. In July 1966 the air force again approached the agency and asked for declassification of the entire Robertson Panel report. But the CIA would not cooperate. Said Karl H. Weber, Deputy Director of OSI, "We are most anxious that further publicity not be given to the information that the panel was sponsored by the CIA."

By the 1970s and continuing into the 1980s, ufologists were convinced the CIA was involved in a massive cover-up and conspiracy. They were in agreement with Dr. J. Allen Hynek who had changed from UFO skeptic to true believer. On November 27, 1978

Dr. Hynek addressed the United Nations on the subject of UFOs. He began by saying:

There exists a global phenomenon the scope and extent of which is not generally recognized It is a phenomenon so strange and foreign to our daily terrestrial mode of thought that it is frequently met by ridicule and derision by persons and organizations unacquainted with the facts. Yet the phenomenon persists; it has not faded away as many of us expected it would when, years ago, we regarded it as a passing fad or whimsy. Instead, it has touched on the lives of an increasing number of people around the world.

He claimed the global phenomenon of UFO reports were frequently made by highly responsible persons. Fifty-seven percent of the U.S. population believed UFOs were real and not a figment of the imagination. Hynek called for the establishment of "a mechanism within the United Nations Organization whereby scientists and other specialists in member nations can bring together and interchange their ideas and their investigative work. .. " He believed UFOs to be eminently worthy of further study and he wanted the UN to help.

In 1977 Donald H. Menzel and Ernest H. Taves published *The UFO Enigma, The Definitive Explanation of the UFO Phenomenon.* One of their criticisms was that too often eyewitnesses would report a UFO sighting, yet nothing would appear on radar. One example they cited occurred in the summer of 1966. A witness was driving in the north central U.S. in a rural area. It was late afternoon. He saw a silvery metallic-looking disk with a dome. It was about 30 feet in diameter and was descending with a wobbly motion into the adjacent valley. It hovered just above the ground about 200 feet from the witness then took off rapidly with a whooshing sound. Physical evidence consisted of depressions in the ground and some overturned rocks. But the nearby radar station at Minot, North Dakota AFB had noticed nothing on its scopes that corresponded to the sighting.

Menzel and Taves tended to dismiss many UFO reports, particularly photographs, as hoaxes. They even included the Trent photos in this category. They were highly critical of witnesses; even pilots and other credible witnesses, complaining most people were not good observers or reporters. They accused the Trents of being "repeaters," meaning they had seen UFOs before, and this somehow disqualified them as reliable witnesses. They were also critical of the Condon Report, but for different reasons than UFO buffs. Forty percent of the University of Colorado Project's cases were unexplained. Menzel and Taves thought this figure way too high; all UFO reports had prosaic explanations in their opinion. Their conclusion: "As a matter of research priority we suggest that all scientists can spend their time to better advantage than in the study of UFOs. Taxpayers' monies, certainly, should not be diverted into this activity. Until and unless evidence of flying saucers of a kind not yet forthcoming appears, our energies should be directed elsewhere."

But UFOs were not going away. During the 1980s the CIA was charged with deliberately withholding information about the 1947 Roswell crash. Finally in September 1994, the air force released a new report that revealed the debris found near Roswell was from a top secret spy balloon called Project Mogul that had been used to monitor the Soviet Union for evidence of nuclear testing. Yet the full truth about UFOs was still being kept from the American public. UFOs might not be the craze they once were, but people were still seeing and reporting strange lights and objects in the skies, particularly odd lighting patterns and black triangles.

A new chapter in the UFO Phenomenon had begun.

3. CLOSE ENCOUNTERS

I cannot presume to describe, however, what UFOs are, because I don't know; but I can establish beyond reasonable doubt that they are not all misperceptions or hoaxes. -- Dr. 1. Allen Hynek

One of the most respected scientists to become involved with UFOs was J. Allen Hynek, a professor of astronomy at Ohio State University. Dubbed "Mr. UFO" by the press, Hynek was one of the few scientists willing to risk his reputation to tackle the UFO problem. Although he began his career as a skeptic, his attitude began to change as he encountered more and more genuinely mystifying cases that seemed to have no ordinary explanations. Looking for natural particularly astronomical explanations for UFOs, he often failed to find them. The more he investigated UFO cases, the more he became convinced there was more to the phenomenon than misidentifications, hallucinations, and hoaxes. As he put it, there were just too many "incredible reports from credible people." He coined the term "close encounters" to classify different categories of UFO sightings, and his book *The UFO Experience* became a classic.

Brought in as the scientific consultant for Project Sign, the first attempt to seriously study UFOs, Hynek would remain with the air force for many years, continuing as the scientific consultant to Project Blue Book. By the 1970s he was no longer with the air force and had left Ohio State to become Chairman of Northwestern University's Astronomy Department. In addition to writing and lecturing, he founded CUFOS (Center for UFO Studies). During his long tenure with the air force he became increasingly frustrated and critical, but he stuck with it in order to have access to its UFO data. Hynek complained that although the Pentagon had begun to treat the subject with "subtle ridicule," it would not open up UFO reports to public examination. The public had no idea how UFOs were investigated nor were they given the results of the investigations. Instead the public was only given the end result, usually in cryptic news releases that left them confused and their questions unanswered. Senior advisers to the Pentagon were not talking about UFOs.

Dr. Donald Menzel, a Harvard astronomy professor, flatly stated that UFOs were nonsense and was a champion proponent of the "mirage theory." His position influenced the Pentagon's position. But if there was nothing to the phenomenon and if the whole idea of UFOs was ridiculous, then why all the secrecy? If the USAF believed UFOs were silly, then why continue a UFO program? Hynek complained that Blue Book investigators "rarely exhibited any scientific interest in the UFO problem." The cases were arranged in folders chronologically without even the most elementary cross-indexing. Hynek was saying more than he probably realized when he asked, "Was all this a smoke-screen, a cover-up job for which Project Blue Book was a front, the real work and information being handled by another agency?"

Scientists ran the risk of professional humiliation if they became involved with UFOs so there was very little exchange of data or cooperation among the scientific community. Hynek thought there should have been a "continuing scientific commission or institute" from the start. The public could have been kept informed in an intelligent and non-sensational manner. Intelligent, competent, and credible people would not have been insulted or made to look foolish, and thus more people would have been willing to assist UFO investigators. International scientific groups could have cooperated on the problem. Lamented Hynek, "It was a mistake from the start to shroud the subject in an air of military science-fiction, an error compounded by seeming duplicity in public press releases and pronouncements."

The following case illustrates Hynek's frustration. On August 1, 1966 a radio station in Cheyenne, Wyoming began receiving calls around 1:30 A.M. about a large, circular object emitting several colors, but making no sound. At the USAF command post near Cheyenne, two officers and one airman controller saw the object directly over the base. It was moving rapidly toward the northeast. At 1:45 A.M., five objects were spotted along with a configuration of two UFOs previously reported over E site at the Sioux Army Depot. At 1:45 A.M., members of E flight saw the same objects, and two security teams were sent to investigate. Reports of sightings continued until 3:40 A.M. One UFO reportedly traveling at a high rate of speed was seen descending and thought to have landed. It was oval and white with white lines on its sides and a flashing red light in the center. When Hynek questioned Major Quintanilla, the head of Project Blue Book, he said, "the sightings were nothing, but stars."

Many of the reports that came to Project Blue Book piqued Hynek's interest because not only were they highly unusual, but highly credible people were reporting them. One such sighting came from a faculty member of the Harvard Medical School and members of his family:

The object caught my attention because ... the light looked wrong for an airplane. We often see, from our house, planes with their landing lights on in an approach to Logan Airport ... There was no sound whatever as the object seemed to get closer ... When the object appeared to be at its nearest point ... a second light appeared on essentially the same course as the first, and my curiosity was further heightened when a third light appeared . .I immediately went indoors for my field glasses. Upon returning, 1 saw that all three lights were still visible, the first two had stopped about 15 to 25 degrees above the horizon and were near to each other and motionless. The third light was still moving. With field glasses, no red, green, or other normal running lights could be seen. At this point the lights came, 1 would guess, about one-half mile apart, a series of them, to a total of six or seven. .. Several of the early lights became completely motionless while others were moving over the horizon: finally two, or perhaps three, of them from the motionless position appeared to drop smaller lights, which twinkled or flashed as they dropped vertically, and as this happened, the motionless lights appeared to dim and extinguish. ... one of the most striking things about the lights was their color. It was orange light and therefore unlike any I have ever seen on an airplane ... too orange to be a normal landing light... the lights continued with no sound that we could detect... the lights were as bright as Venus as seen at its brightest ... but they certainly cast no light on the ground . . I could see no shape or form of anything else attached to these lights.

Another case in Hynek's files came from a professional astronomer who waited eleven years to report his sighting because he feared ridicule:

We followed country roads until we came within 100 yards of the object. It was hovering around a large tree, which stood alone in the center of a cultivated field .. The object, which subtended an angle of about 1/4 degrees (giving it a physical diameter of less than 3 feet), appeared circular in shape and was thus probably a spheroid It was highly luminous against the dark sky background and changed color through the whole visible spectral range ... The object appeared to be examining the tree rather closely. It circled the upper branches, ranging from 50 to 100 feet off the ground, passing in front of the tree, then clearly visible, through the branches on passing behind the tree again. It continued this apparent "observation" of the tree for several minutes while we watched Then, anxious for a picture, we climbed the perimeter fence and started slowly toward the tree facing due west. We had not gone more than 10 feet before it "noticed" us and, noiselessly accelerating at a very high rate, headed almost directly south, disappearing over the horizon (on a slightly rising trajectory) in about 2 1/2 seconds.

This case is an example of what Hynek called Close Encounters of the First Kind: UFO sightings of objects or bright lights close to the observers, generally less that 500 feet away. There is no interaction between the UFO, the witness, or the environment.

A police officer in North Dakota reported an object he saw at 3 :00 A.M. It hovered before moving away rapidly: "When I sat there, I had a sort of fear; it wasn't for myself, but for what it might mean. I sat there, I suppose for about five minutes. It bounced up and down ... but when it left, it was gone -- **bang** --it was out of sight in less than five seconds .. .In my mind, it was guided by somebody or something."

A Close Encounter of the Second Kind is a UFO encounter where there is some kind of physical effect, where a visible record of the UFO has been left. Marks on the ground, scorching or blighting of vegetation, unusual or anxious animal behavior, car engines that temporarily stop, a person feeling heat, car headlights that dim or go out would all be examples of this type of Close Encounter. One such case comes from a 1967 encounter in Massachusetts:

When we got close to the object, the car stalled and the lights and our radio all went off at the same time. After this, I tried to start the car twice while the object appeared to remain stationary .... I shut the light switch and the radio off. Then I tried to start the car again twice. It did not start. Next, the object in the sky seemed to start moving away from us. I tried to start the car again, and it immediately started .. .I saw an object to the left of us in the sky, which at first appeared to be a plane. As we approached it, I saw that it was too large and too low to be a plane and called the attention of the other occupants to it ... The object was moving in the same direction as we were at first, then stopped for about a minute, then flew off, and the car started again. The object made no noise ...

In Quebec on May II, 1969 M. Chaput, a pulp mill worker, was awakened on his farm at 2:00 A.M. by his barking dog. Looking out, he saw an intense light illuminating his field. It lit the house as well as the surrounding area. The light then moved off silently although he heard a "purring sound" as it receded into the distance. The next morning he found a circular mark, equally spaced to form a triangle with a rectangular depression that was one to two inches deep.

On March 20,1992 at about 3:50 in the morning patrolman Luis Delgado in Haines City, Florida was in his patrol car when he saw a green light in his rear view mirror. Suddenly a green glow filled the interior of his car. The light continued to follow the vehicle and Delgado reported this to Police Dispatch and asked for backup. He pulled off the road. Then his engine, lights and radio stopped. Delgado later said the object was about 15 feet long, thin, with a three-foot high center. It was a strange green color that "seemed to flow over the surface." The object hovered about 10 feet off the ground.

While he watched it, the UFO shone a bright white light into his patrol car. Delgado jumped out and tried to call Police Dispatch on his walkie-talkie. The air temperature around him dropped and he said he could see his breath although official records show the actual temperature at that time was 60 degrees. Just as quickly as the object appeared, it suddenly left, speeding off in a few seconds, staying low to the ground. When another police officer finally arrived, he found Delgado sitting in his car, shaking and crying and unable to talk. After he recovered, Delgado filed an incident report. His car returned to normal and he himself suffered no side effects from the encounter.

The most dramatic encounter with a UFO is a Close Encounter of the Third Kind, the kind portrayed in the movie by the same name and starring Richard Dreyfuss. Hynek described this type of encounter as the presence of "animated creatures," usually described as humanoid. Although a Close Encounter of the Third Kind can generate the most laughs and jokes about LGM (Little Green Men) or BBW (Big Blue Women), Hynek felt there were too many of these encounters to simply dismiss them. His colleague, Jacques Vallee, prepared a catalog of 1,247 cases. Of those, 750 were of an actual landing of a craft and more than 300 (about 40 percent) were reported to have humanoids in or near the craft. Blue Book tended to dismiss these cases, and their files had only 48 landings with only 12 humanoids.

A major problem with this type of close encounter is its association with the contactee, transportee, and abductee movements which serious UFO investigators generally dismiss as too wildly fantastic to be taken seriously.

Another problem is the presence of humanoids. If these beings are truly from outer space, then why do they resemble us? How is it they can breathe our air and adapt to our gravity and air pressure? This too bothered Hynek: "Something seems terribly wrong about that. This would imply that they must be from a place --another planet -- very much like our own. Perhaps our own? But how? Or are they robots, not needing to adapt to our environment? Our common sense recoils at the very idea of humanoids ... They tend to throw the whole UFO concept into disrepute."

Witnesses never see silicon-based giraffe-like creatures or low-gravity octopus-like aliens. The creatures are always human-like all of the time. This is a very important point. One of the first reports of a close encounter with an alien being took place on August 21, 1955 in Kentucky.

A UFO landed in a gully near the farmhouse belonging to a family named Sutton. One witness saw the craft land and told the rest of the family. No one believed him and they ridiculed him. But then the dog in the yard started barking violently. Two men went to the back door to check on the dog and saw "a small glowing with extremely large eyes, his arms extended over his head as though he were being robbed," slowly approaching the house. One man grabbed his .22 rifle and the other got his shotgun. When the being was about 20 feet from the house both men fired, the sound described as 'lust like I'd shot into a bucket." The being did a flip and ran away.

Another being appeared at the window. Again the men fired. The family went outside to see if the alien had been killed. As the first man stopped under a roof overhang, the others saw a clawlike hand reach down from the roof and touch his hair. The men fired at the being on the roof and at another one they spotted on a nearby tree branch. This one was hit directly and "floated to the ground and scurried away."

Highly agitated, the Suttons barricaded themselves inside their house and bolted the doors. But the aliens kept appearing at the windows. After three hours of this activity, the eleven-member family piled into two cars and headed into town to notify the police. The terrified Suttons returned with the police who found nothing despite an extensive search.

After the police departed, the creatures came back. Bud Ledwith, an engineer and announcer for WHOP in Hopkinsville, investigated the case and gave his files to Hynek. Ledwith questioned all seven adult witnesses and asked them to draw the creatures to the best of their recollections. Signed statements were then obtained from each witness. Ledwith also made a composite drawing of the beings and the witnesses signed it. Eventually this is the picture that emerged:

The eyes were like saucers, large and set about six inches apart; they seemed to be halfway around the side of the face ... The head itself was circular and completely bald on top ... We progressed to the body. No one was sure whether there was a neck or not, so we left it out. According to the women, the body was thin, with a formless straight figure ... The arms were peculiar; they were almost twice as long as the legs ... the hands were huge, bulky looking things ... The only part of the face that no one could describe was the nose ...I tried to sketch in a nose ... but no one was sure so we removed it.

Later, Ledwith also interviewed the Sutton men, especially Lucky Sutton who was the household's patriarch. Lucky said, "No, the face is almost round, it doesn't come to a point." Lucky claimed there was no mouth but if there was, it was just a straight line from ear to ear. What impressed Ledwith was that all seven people gave him the same story and nearly identical pictures. Since the family members had not consulted with each other, he felt certain that their accounts were honest and sincere. " ... all seven were sure of what they had seen, and no one would retract a statement. .. even under close cross-examination."

Ledwith also said he was fortunate to interview the family before the "case received so much adverse publicity and personal harassment that they soon refused to discuss the matter with anyone."

However, one year later, Isabel L. Davis of New York City was able to interview the family again. She too shared her conclusions with Hynek. Like Ledwith, she was certain the Suttons were being truthful and honest. " ... the Suttons stuck to their story. Stubbornly, angrily, they insisted they were telling the truth. Neither adults nor children so much as hinted at the possibility of a lie or mistake -- in public or to relatives; there was no trace of retraction."

Hynek felt the accounts given to Ledwith and Davis "give us a picture of truly bizarre and, in ordinary terms, completely unexplainable event. Seven adults and four children attested to the essentials of the event." Hynek also personally knew both Davis and Ledwith. Ledwith had been in his employment for two years on a satellite-tracking project that involved setting up twelve satellite tracking stations around the world. Ledwith had worked on the crystal-clock timing mechanisms of the Baker-Nunn satellite cameras. Blue Book made little or no investigation on the case and their records of the incident are sketchy. Yet in the files it is listed as "Unidentified."

Hynek noticed that the "strangeness spectrum" of UFO reports is narrow and restricted. In other words, UFOs do not run the gamut of all possible configurations. The vast majority of credible reports are essentially consistent with one another. The objects are invariably described as triangular, oval-shaped or circular. They are metallic. They hover or remain motionless and then fly off in great bursts of speed; they are noiseless or make a "whooshing" sound. There is no sonic boom or other loud noises. Their actions appear to be controlled, although they are sometimes seen wobbling, tumbling, or falling. They exhibit extremely rapid acceleration that seems to violate the laws of physics. They leave vapor trail or exhaust, they appear as bright objects even in daylight, they have odd lighting patterns. When UFO occupants are seen, they are always human-like in form.

This uniformity among sightings and reports led Hynek to conclude that UFOs, whatever they are, must be real objects and not the products of imagination or delusion. He strongly disagreed with the majority of his fellow scientists who either regarded UFOs with contempt and dismissed them out of hand or saw them as a purely psychological phenomenon. For Hynek UFOs were worthy of further study.

In addition to visual sightings, UFOs are occasionally picked up on radar. "The speeds involved are invariably high, but combinations of high speed at one time and hovering at another are not uncommon. Reversals of motion and sharp turns ... are characteristics of Radar-Visual cases." One of the most famous cases was the Washington D.C. Saucer Flap in 1952. Although the air force dismissed the sightings as "temperature inversions," the highly unusual blips prompted the formation of the Robertson panel, a committee of five distinguished scientists who spent five days the following year investigating UFO reports.

Another famous R-V case occurred on November 4, 1957 in New Mexico:

... both sources were on duty alone in the control tower at Kirkland Air Force Base, New Mexico ... One of the controllers looked up to check cloud conditions and noticed a white light traveling east between 150 and 200 miles per hour at an altitude of approximately 1500 feet ... Source then called the radar station and asked for an identification of the object. The radar operator reported that the object was on an approximate 90-degree azimuth from the observer; it disappeared on 180-degree azimuth from the tower observer. The object angled across the east end of runway 26 in a southwesterly direction and began a sharp descent. One Source gave a radio call in an attempt to contact what was believed to be an unknown aircraft that had become confused about a landing pattern ... The object was then observed through binoculars and appeared to have the shape of "an automobile on end" This was estimated to be 15 to 18 feet high. One white light was observed at the lower side of the object. the object slowed to an estimated speed of 50 miles per hour and disappeared behind a fence at "Drumhead, " a restricted area .. .It reappeared moving eastward and one Source gave it a green light from the tower, thinking it might be a helicopter in distress. The object at this point was at an altitude of 200 to 300 feet; it then veered in a southeasterly direction, ascended abruptly at an estimated rate of climb of 4,500 feet per minute, and disappeared ..

About 20 minutes later, the object was again picked up on radar. This time it was tracked for 20 minutes before it faded from the scope. During that time, it flew at a high rate of speed, made an abrupt turn and fell into trail formation with an AF C-46, turned north and hovered for one and a half minutes.

Hynek didn't know what to make of this case. He was sure the object was real, but how could it look like a car standing on end? It couldn't have been an ordinary aircraft because the two radar operators with 23 years of control tower experience between them couldn't identify it. Although Blue Book did no follow-up and made no detailed investigation, it concluded the object was a "misguided aircraft."

. On the afternoon of July 17, 1957 an air force RB-47 was followed by a UFO for 700 miles as if flew over Mississippi, Louisiana, Texas, and Oklahoma. The cockpit crew saw the object as an "intensely luminous white light." It was also detected by ground radar and ECM monitoring gear on the RB-47. The crew noted the rapidity of its maneuvers which none of them had ever seen before.

A signalman on a navy ship in the Philippines reported what he believed to be an aircraft. Viewed through binoculars, the aircraft turned out to be three different objects which were "traveling at extremely high speed." Although three objects were reported visually, air search radar detected four moving targets and held them for six minutes. When they were directly over the ship, the objects spread out in a circular formation and remained there hovering for three minutes. This maneuver was observed visually and detected by radar. A bright object that hovered off starboard made a larger image on the radarscope. The objects were tracked at speeds in excess of 3,000 knots before moving off in a southeast direction. Blue Book dismissed this 1965 case as having "insufficient data."

R-V cases tend to be among the most disturbing, because they involved highly credible and highly trained observers who are accustomed to seeing and tracking a wide variety of targets. Cases that involve both visual reports and radar confirmation are especially troublesome for here is "hard data" about UFOs.

On May 4, 1966 an air traffic controller in Charleston, West Virginia saw a target to the left of Braniff Airlines Flight 42. The operator had been an air traffic controller for 13 years, three in the air force and ten with the FAA. He was working the high altitude radar sector on the midnight shift. At 04:30 he saw the target. He advised Braniff he had no known traffic in his vicinity, but was "painting a raw target off his 10 o'clock position." He asked Braniff for a description of the object. Braniff 42 said it "was not an aircraft, the object was giving off brilliant flaming light consisting of alternating white, green, and red colors ... " The radar operator said the target was "doing approximately 1,000 mph and made a complete 180-degree turn in the space of five miles, which no aircraft I have ever followed on radar could possibly do, and I have followed B-58s declaring they are going supersonic ... "

One of the more unusual R-V cases occurred on January 13, 1967 in Winslow, Arizona. The object was spotted at 10:00 P.M. and was observed for 25 minutes. The pilot of a Lear jet reported a red light at his 10 o'clock position. The light flashed on and off and then quadrupled itself in a vertical direction. Albuquerque radar painted an object when the light was on, but none when it was off. The light repeated its quadrupling process several times, "seeming to retract into itself the lights below the original light." The Tower warned the jet that the object was getting closer. It then accelerated rapidly and seemed to be playing a cat-and-mouse game with the jet. After "some 25 minutes and with terrific acceleration ... the object ascended at a 30 degree angle and was gone in fewer than ten seconds." Albuquerque continued to track the object until its final acceleration and disappearance.

Hynek couldn't understand the air force and Blue Book's cavalier attitude toward UFOs. The more he investigated reports and sightings, the more he became convinced UFOs were real and needed to be studied scientifically. He yearned for a statistical analysis of UFO cases in order to identify data patterns and the measure of repeatability. He thought such a study would yield "scientific paydirt" and the results would mark a "totally unexpected quantum jump" in science. He thought UFOs represented something totally new to science and their study could only enrich humankind.

4. CHAPTER FOUR

The Flying Triangles

When presented with two theories that explain the data equally well, choose the simpler theory. -- Occam's Razor

It's us. What else could it be? - Jim Penniston, USAF Technical Sgt., Retired, a witness to the Rendlesham Forest Incident

You don't know the half of it. -- Former President George Bush I responding to a White House staffer who asked him how much he knew about UFOs

Boomerang or triangle-shaped UFOs have been seen for many years in both the U.S. and the UK. Sightings of flying triangles in the U.S. were reported as early as 1951 and then continued sporadically throughout the decade. In the 1960s one of the most famous incidents was the Ann Arbor, Michigan UFO flap when Dr. J. Allen Hynek had his close encounter with swamp gas.

During March 1966 there was a well-publicized cluster of UFO sightings near Ann Arbor. One deputy sheriff reported on the objects' "fantastic speeds," "sharp turns," and "how they could dive and climb and hover with great maneuverability." Another witness described the UFO as " sort of shaped like a pyramid (read triangle) with a blue-green light on the right hand side and on the left a white light." He continued, "You couldn't see it too good, because it was surrounded with heat waves, like you see on the desert. The white light turned to blood red as we got close to it." Then the object disappeared.

More than 50 people, including several police officers, reported seeing the object apparently hovering over a swamp. On their way back to Ann Arbor, the police in one squad car saw the UFO and chased it at high speed, but to no avail. The next day another 50 people, including 12 policemen, saw the same pyramid like object. That same evening, students at nearby Hillsdale College, a local civil defense director, and a college dean saw a mysterious object flying around and flashing bright lights in a swampland.

The whole Ann Arbor area, as well as the state of Michigan, was in an uproar. Wide coverage by the national news media added to the frenzied, hysterical atmosphere. Pressure increased on the air force to do something, so Dr. Hynek was dispatched to investigate. Pressed into making a statement before he had drawn any clear conclusion, he offered his now legendary "burning swamp gas" explanation.

In retrospect it's easy to see that Hynek and many other experts had missed the most essential and important facts about this incident. The object 'had been described as pyramid-shaped which would indicate a triangle or delta shape and the objects had flown at extremely high speeds as well as executing fantastic maneuvers.

In the mid-I 960s, in Charlton Park in Great Britain, a flying triangular-shaped object was spotted. A woman was walking through the park when she saw a triangular craft that made no sound at all. When she came out of the park, she saw it again. This time it made a whooshing noise before it suddenly disappeared.

These two incidents, although in different countries, were among the increasing reports by witnesses of odd, triangular shaped craft that made an odd sound or no sound at all, and could fly as well as maneuver at high speeds, Flying triangles would be responsible for a large number of UFO sightings over the following years.

In August 1972 two teenage boys "Paul" and "Ed" went on a camping trip with Ed's family. Both boys were 18 and had just graduated from high school. Greg Long describes their experience on his website "The Triangle and the Road Taken." They had set up their campsite near the Carbonado Entrance to Mount Rainier National Park:

The triangle of lights hurtled through the air in the northeastern quadrant of the sky. Within that circumscribed space, the triangle flew in no set pattern, its motion like a pinball let loose to impulsively bounce and careen, always in straight lines. Each line of travel lasted perhaps half a second, although long lines of travel that covered larger distances took a bit longer ...

Gradually, the triangle made its way south, crisscrossing over its own lines They set up tents near the Carbonado entrance to Mount Rainier National Park. That night, past midnight, Paul and Ed lay down in the back of the station wagon, their heads next to the tailgate. The window in the tailgate was down, and they could see, there above, stars beyond measure, twinkling and glowing in the deep, oceanic expanse of the night ... Then perhaps 5 to 15 minutes later, they showed up. They were lights, three of them. They were about the size of stars. And they were green, red, and white. if one wasn't careful, one would think they were airplane lights. The lights were arranged in the form of a triangle, one light on each of the three points of the triangle. Ed said, "'That's a plane, right?" The triangle of three lights was moving rapidly from the young men's right, from the north, and was flying at breakneck speed Suddenly it shot upward, as if ricocheting off an invisible wall. Then it shot to the right, down, left, up, right, down, left, scribing a series of straight lines and rectilinear pal/ems. The angular turns were abrupt, very sudden, cutting 90-degrees each time. "No," Paul said, "I don't think we have anything that flies like that. " from Greg Long's The Triangle and the Road Taken.

The triangle of lights was described as flying faster than a plane but silent. The speed of the object coupled with its silence convinced the boys it was not from Earth. Remember the Mount Rainier area was where Ken Arnold had made his original sighting in 1947. Also note that the lights went "left, up, right, down, left ... "

One of the most notorious flying triangle episodes comes from the UK. This bizarre incident which occurred over three consecutive nights in late December 1980 is likened to the Roswell incident in the U.S. Two USAF security policemen saw strange lights in Rendlesham Forest outside the back (east) gate of RAF Woodbridge. RAF Bentwaters and RAF Woodbridge were twin USAF and RAF bases near Ipswich in Suffolk. A fourteen square mile patch of woods called Rendlesham Forest separates the two bases. At the time of the incident, the bases were home to the USAF 81 st Tactical Fighter Wing, the largest tactical fighter wing in the free world during the Cold War. RAF Woodbridge also contained one of NATO's largest secret stockpiles of nuclear ordnance.

At first, the two men thought the unusual lights belonged to a downed aircraft. They received permission to go into the forest and investigate. A third patrolman joined them. The three men saw a strange, glowing, triangular object maneuvering through the trees. On top of the object was a pulsing red light and beneath it were blue lights.

One of the patrolmen was USAF Technical Sergeant Jim Penniston, now retired. Penniston was trained to identify all types of aircraft, and in those early morning hours of December 27, 1980 he took a camera and notebook with him. In 1996 Penniston gave an interview to *Omni* magazine contributing editor A.J.S. Rayl. The following is excerpts from that interview.

Penniston told Rayl at about 12:02 A.M. he was dispatched to the east gate of RAF Woodbridge. When he arrived, Staff Sgt. Steffans told him there was a problem in the woods. Penniston saw what appeared to be a fire about 300 meters away and figured it was an airplane crash. But Sgt. Steffans said the aircraft must have landed because there had been no sound of a crash. Nevertheless, Penniston still thought it was a downed aircraft because he could see orange, red, and blue glowing lights. From the gate shack, he phoned his superiors at Bentwaters Center of Security Control (CSC).

Instructed to stand by while the Control Towers at Bentwaters and Woodbridge were contacted, he was next told a bogie (UFO) had been reported about three miles out from Bentwaters, but then contact had been lost. Still thinking the lights were from an aircraft crash, he asked for and received permission to go off base to investigate. Two other security policemen, Airman First Class John Burroughs and Airman First Class Ed Cabansag, went with him. Before setting off into Rendlesham Forest, they were told to leave their weapons behind.

The three men got into a jeep and drove out of east gate and down a logging road. The road got too rocky and they stopped the jeep. Because their radios were experiencing a dampening effect, Cabansag was stationed as a radio link while Penniston and Burroughs headed on foot toward the forest. Approximately 50 meters away was a clearing (Capel Green) from which the lights appeared to be coming.

As the two patrolmen got closer, Penniston said, "It was apparent that it was not an aircraft downing or a crash." There was no smell of burning fuel, and the men could feel electricity in the air.

They saw an object sitting in the clearing, emitting a very bright white light. The large, glowing light was moving left--right, left--right (see the Greg Long account). The light was so bright that both men had to squint when they looked at it. The object made no sound, "but the animals around us were in a frenzy."

Penniston left Burroughs behind as a radio link while he continued into the woods. The lights which had been a blur were now definite, distinct colors. He clearly saw blue. yellow and red pulsating lights. And the object was definitely not a conventional aircraft. "It was no aircraft I had ever seen, and it wasn't one that I knew any prototype of." He wrote in his notebook:

Triangular in shape. The top portion is producing mainly white light, which encompasses most of the upper section of the craft. A small amount of white light peers out the bottom. At the left side center is a bluish light, and on the other side, red. The lights seem to be molded as part of the exterior of the structure, smooth, slowly fading into the rest of the outside of the structure, gradually molding into the fabric of the craft.

Then he took out his camera and "began snapping photo after photo. It was still eerily quiet." He took all 36 pictures on his roll of film.

He estimated the craft was about three meters tall and about three meters wide at the base. Although he didn't see any landing gear, he thought it seemed like "she was on fixed legs." He walked around the craft and then right up to it. There were no access ports or windows. The surface was completely smooth. Static electricity was still in the air. He said, 'The fabric of the shell was more like a smooth, opaque, black glass. The bluish lights went from black to gray to blue. I was pretty much confused at this point...It was dead silent. No animals were even making noise anymore."

On the smooth exterior of the shell there was writing of some kind, but I couldn't quite distinguish it, so I moved up to it. It was three-inch lettering, rather symbols that stretched for the length of two feet, maybe a little more. I touched the symbols, and I could feel the shapes as if they were inscribed or etched or engraved, like a diamond cut on glass.

    At that point, I backed away from the craft, because the light was starting to get brighter. Still, there was no sound. There was no physical contact with any kind of life form, but there did seem to be a life presence. It was mechanical, this ship, and it seemed to be under intelligent control.

    The next thing I knew, I was standing about 20 feet away from the craft with Burroughs, who I thought I had left back near the tree line. The craft moved up off the ground, about three feet, still with absolutely no sound. It started to move slowly, weaving back through the trees at a very slow pace, maybe half a foot per second. It took about a couple of minutes for it to maneuver itself back to a distance of about 100 to 150 feet, then it rose up just over the trees, about 200 feet high. There was a momentary pause and then literally with the blink of an eye it was gone. All with no sound. That still boggles my mind.

After the object disappeared, Penniston and Burroughs saw "the same array of colored lights maybe a half mile away." They tried to pursue it, but decided to turn around and go back to the clearing. There, they found three indentations in the ground, all triangular in shape and each about three meters apart. Later, during their debriefing by the assistant operations officer, they were asked to explain what happened:

    So we explained, very briefly. We didn't tell him any specifics about the symbols or the design of the craft. After a very long pause and very calmly, while he was tapping his pencil on the table, he said, "Gentlemen, what you say you experienced tonight is no longer able to be reported through Air Force channels. "He then gave us something of a history lesson on Project Blue Book and that it was terminated in 1969. Basically, he told us there was no official way to report this up. Then he said, "Some things are best left unsaid. " He asked us to keep quiet about it, to forget it happened.

Penniston and Burroughs were given a six day authorized break and they went home to Ipswich. Penniston dropped Burroughs off and then went home. He changed clothes, got some plaster, and drove back to Woodbridge. He went back to the Capel Green clearing where they had seen the indentations. He poured plaster into them and made plaster casts of each impression.

He never got his roll of film back. "I had dropped my film off at the base lab for developing, but I never got it back. I never saw them. I was just told that they didn't turn out."

In 1984 Penniston was assigned to Grissom AFB in Indiana. He said he found a listening device in his house on the base. He and Burroughs both had harassing phone calls. Once, during a burglary, Burroughs' video documentation of the incident and his files on Bentwaters were the only items taken. Penniston put it all down to "damage control." Unfortunately, this was one of the few times that actual physical documentation and evidence could support a UFO encounter. And what of the plaster casts he made? Cast #1 was included with his family's belongings for the shipment home. He hand-carried the other two on the airplane when he returned to the US. When they received the shipment (after much delay) everything was there but the cast. He buried #2 and #3 "to insure that they wouldn't go missing." Penniston showed one on the Sci Fi channel's special about Rendlesham. He told the Sci Fi channel that although he doesn't believe in aliens, he thought the Rendlesham UFO could indicate some kind of time travel from our own future.

But that was not the end of the Rendlesham encounter. The next night, December 28, 1980 Lieutenant Colonel Charles Halt, the Deputy Base Commander, was at a holiday party when a white faced, on duty, flight commander came in and announced, "It's back." Lt. Colonel Halt and a team of officers went out to investigate. They had with them night vision scopes, Geiger counters, and cameras. He too was interviewed by *Omni* contributing editor A.J.S. Ray!:

J took half a dozen of the men and headed into the woods onf oot to a clearing where the initial incident had supposedly taken place. (It later turned out that Penniston and Halt had actually been at two separate locations in the forest, something they didn't know until years later.) *We found three distinct indentations in the ground equidistant apart and pressed well into the sandy soil* (Bentwaters and Woodbridge are near the North Sea). *They were supposedly caused by the object seen two nights before,* (actually the previous night) *but J didn't see anything sitting there that night .. ...Inside the triangular area formed by the indentations, one of the men got slightly higher readings on the Geiger counter than he did outside. He photographed the area, and J took a soil sample. Meanwhile, J recorded this activity on my micro-cassette recorder.* (The triangular impressions indicate Halt either saw the same object as Pennistion or one very similar to it.)

Suddenly, toward the east, Halt and his team saw "an unusual red, sun like light -- oval shaped, glowing, with a black center -- 10 to 15 feet off the ground, moving through the trees .. .The animals were making a lot of noise." The object moved in a 20 to 30 degree horizontal arc and seemed to be "dripping what looked like molten steel." Then the object exploded. But "not a loud bang, just booompf -- and broke into five white objects that scattered in the sky."

In the north the team spotted three white objects. These objects were elliptical with blue, green and red lights on them. The objects made "sharp, angular movements and eventually turned from elliptical to round." Halt called the command post to see if anything was appearing on radar. They told him that they didn't see anything. They repeated the statement:

Suddenly, from the south, a different glowing object moved toward us at a high rate of speed, came within several hundred feet, and then stopped A pencil-like beam, six to eight inches in diameter, shot from this thing right down by our feet. Seconds later, the object rose and disappeared The objects in the north were still dancing in the sky. After an hour or so, I finally made the call to go in. We left those things out there.

The film turned out to be jogged; nothing came out. But a staff sergeant later made plaster castings of the indentations, and I had the soil sample. Around New Year's Eve, I took statements and interviewed the men who had taken part in the initial incident. The reports were nearly identical.

On January 13, 1981 Halt wrote a memo to the British Ministry of Defense (MoD). In 1983 Halt's memo was released in the U.S. through the Freedom of Information Act. Although he wrote a sanitized version, it still caused a sensation in UFO circles (See the last page for the complete memo).

On the night of December 29, 1980 eighteen-year old Airman First Class Larry Warren, a member of the 81st USAF security police, was on duty. Around 1:30 A.M. an army truck pulled up and he was ordered inside. He and other air force personnel were driven from the Bentwaters base toward the Woodbridge base. At the east gate, they made a left turn and then stopped near Rendlesham Forest where they got out of the truck and handed in their M 16s.

Staying in small groups, they moved into the woods where they saw a yellowish-white mist about 50 feet in diameter and about two feet high over the clearing called Capel Green. They also saw a red light moving erratically above them. Suddenly the light approached the clearing at a fantastic speed. It emitted a blinding flash of light as it touched down on the mist. The soldiers later said they had to cover their eyes because the light was so bright. When they opened their eyes, they saw a large triangular craft sitting in the mist.

Warren and the other soldiers were ordered to surround the craft. Meanwhile, some civilians who had also spotted the strange object had arrived and were taking pictures and videos. A shaft of light shone from the triangle and there were three beings visible inside the craft. Warren thought the beings had large heads, dark eyes and appeared to be semi-transparent. The beings emerged from the triangle and approached Colonel Gordon Williams, the wing commander on the scene. The beings and the commander seemed to engage in a telepathic conversation. After about half an hour, the craft vanished. "It was gone in a flash," said one witness, "almost like it just disappeared. When it left, we were hit by a cold blast of wind which blew toward us for five or ten seconds. It was a really scary feeling ... "

Warren and the others were told to return to the place where they had left their M16s. When they got there, a smaller craft' was hovering over the trucks. Some of the soldiers grabbed on to this craft and were lifted from the ground and flown around a bit before they let go.

Neither Airman First Class Warren, nor the other soldiers

spoke of the encounter until the next day. The British police film of the incident was confiscated and the men, after a thorough debriefing, were ordered not to talk about it. But Warren telephoned his mother back in the United States and started to tell her about the strange incident. As he talked, the phone suddenly went dead. Military officials later approached him with tapes of his conversation with his mother and then fined him for violating the -talking order. Many of the other soldiers involved in the incident were sent to Egypt and elsewhere to "cool down." Warren was given a job in the offices. While he was there, he tried to learn everything he could about that night.

Unlike Penniston and Halt, Warren described a Close Encounter of the Third Kind where contact was made with beings inside the craft. Colonel Williams, Commander 81st Tactical Fighter Wing, dismissed Warren's account.

According to *UFO Invasion at Rendlesham,* a Sci Fi Channel Declassified Special, Larry Warren and other witnesses were taken to a secured facility where they were made to watch films of UFOs and sign documents. They were supposedly told "bullets are cheap" in order to intimidate them. Warren later claimed he had post traumatic stress. There were charges of brain control and mind altering drugs.

No official information was ever released about Rendlesham. The incident was suppressed under the American National Security Act of 1947 and Her Majesty's Official Secrets Act. Then in October 1983 the British press exploded with Larry Warren's eyewitness account. By the end of the month both houses of Parliament held a closed-door session on the subject. Both the British Ministry of Defense and the U.S. Department of Defense, while admitting the incident did happen, declared that it was not a threat to the security of either nation, nor was it a threat to any military bases in Britain.

In 1987 Larry Warren asked Peter Robbins to write a book with him about the incident. The two men spent nine years investigating the Rendlesham Forest Incident. The book that finally resulted was titled *Left at East Gate.* Peter Robbins wrote on his web site: "It is the nature of governments to keep certain things from their people, either for the public good or to maintain official control. But the process of enforcing secrecy results in the gradual erosion of

61

democratic institutions in the name of national security. In the Bentwaters case, truth and reason have been the casualties for the citizens of the United States and Great Britain. How much longer will government secrecy continue to confound the UFO issue?"

Halt was never formally debriefed. Military officials claimed a satellite had fragmented as it reentered the Earth's atmosphere. Another explanation offered was a refracted light beam from a nearby lighthouse. Rabbit burrows caused the triangular depressions. The numerous lights witnesses had seen were stars or meteorites. The USAF has so far refused to comment on the incident while the British MoD maintains it was no threat to the security of Great Britain.

One person who did not accept these explanations was Lord Hill-Norton, Admiral of the Fleet and Chief of the Defense Staff from 1971 to 1973. So far, he is the highest-ranking military figure to go on record about Rendlesham. He believes that something actually landed because the witnesses were military people who were trained to observe. Convinced there had been a cover-up, Lord Hill-Norton made inquiries in 1985. He was told, "The events to which you refer were of no defense significance."

This too was unacceptable to the Former Admiral of the Fleet. Either the events described in the Halt memo actually took place or else Halt and his team were either lying, delusional, or mad. Each scenario was equally dangerous in Lord Hill-Norton's view. It either meant a UFO had penetrated British airspace or a bunch of kooks held important posts at one of Britain's most important bases, a base that held a large, secret nuclear stockpile. Either way, he felt the incident was of "extreme defense significance. "

The Rendlesham Forest Incident was not the first time a UFO had been spotted in this area of Ipswich in Suffolk. On August 13, 1956 at 9:30 PM radar operators at the Bentwaters base detected an object about 30 miles out over the North Sea. This is Case #2 in the Condon Report, and contains a detailed account of the incident by the Watch Supervisor at the GCA station in the Radar Air Traffic Control Center. The committee also received the Project Blue Book files on the case.

The object was heading inland at an incredible speed, covering six miles with every four-second sweep of the radarscope. The controllers estimated its speed at almost 5,000 mph. Meanwhile, other objects were also being picked up on the scope and were moving toward Bentwaters: three objects in a triangular formation were preceding a cluster of between twelve and fifteen echoes. The pilot of a Lockheed T -33 jet trainer, who was also flying toward Bentwaters, was asked by the controllers to investigate, but he saw nothing.

As the controllers watched their scopes, the cluster continued toward Bentwaters and then at a range of about 40 miles it merged to form a single, powerful radar echo. The echo remained stationary for some fifteen minutes, then moved to the northeast, stopped again for a few minutes, then gathered speed and vanished from the radarscope.

Meanwhile, the original, single echo had been lost, but at 10:00 PM another blip (or maybe the same one) appeared on the Bentwaters radarscope moving at an estimated 4,000 mph across the screen. The strange object was also detected on radar at nearby Lakenheath, another USAF base in Suffolk. A RAF Venom NF-2 was scrambled to intercept the UFO. At one point, the object was stationary and the Venom was soon within gun range. But the radar operator lost contact because the UFO had moved behind the Venom crew and was keeping pace with them. For the next ten minutes, the Venom pilot tried to shake off the UFO, but it continued to follow them, keeping at a distance of about 200 yards. Eventually the pilot managed to break away and headed back toward base. The UFO followed for a while, stopped, then headed north at 600 mph until it vanished from the radar screens. A second Venom was scrambled, but it was unable to intercept the UFO.

The Condon Report concluded that while a conventional or rational explanation could not be ruled out, it seemed likely "that at least one genuine UFO was involved." J. Allen Hynek thought this was certainly one UFO case worthy of further study. But others put the unusual radar sightings down to "anomalous propagation."

Another well-documented case of flying triangles occurred in Belgium. On November 29, 1989 a UFO flap began in that country. Most of the sightings were in the region of Wallonia where thousands of people reported seeing triangular shaped aircraft in the skies. Other sightings occurred on December 15 near Liege and on March 30-31 in the area of Wavre, Tublize and Brussels. The triangles flew in such an unusual way that people were convinced they were not terrestrial aircraft. Air traffic controllers, police officers, and personnel in the Belgian Air Force confirmed these civilian reports.

Colonel de Brouwer of the Belgian Air Force said some of the air force's tracking data showed the craft making "extreme, instantaneous changes in velocity and altitude. Calculations of some of the sighting data indicates gravitational stress in the 40G range. Needless to say, this is quite impossible from accepted thinking in physics for any 'manned' aircraft. We believe it is safe to say that any human occupant aboard an aircraft performing such actions would be little more than a wet spot on the wall afterward."

The maneuvers executed by the craft were "done at altitudes virtually impossible for the F-16 interceptors launched for investigation to duplicate." The objects also flew and maneuvered at speeds which "definitely broke the sound barrier, but with no shock wave registering, and no sonic boom being heard by ground observers. "

Another alarming aspect to these triangles was their apparent working in unison with each other. They would move slowly "over certain countryside areas, hovering, and sweeping the ground with what appear to be either laser beams or high intensity lights." The conclusion was made that these triangles could not be explained by "ordinary aircraft technology." The possibility that they might be secret military aircraft was also ruled out as "the logical explanation."

The Belgian Air Force scrambled F-16 fighters in an attempt to intercept the UFOs. Radar returns from different stations showed the UFOs accelerating away from the F-16s at unbelievable speeds and performing maneuvers that amounted to g turns. Ground witnesses, including policemen who saw the UFO in November, said it was a large, triangular object that was hovering over a field. Three bright lights could be seen on the underside of each apex with a red-orange light in the middle. The object moved toward the men and they heard a humming noise as it flew over them.

Many UFO buffs were convinced that the Belgian flap was incontrovertible proof of the extraterrestrial origin and explanation of UFOs. But the skeptics denied this and put the sightings down to bright stars and planets and radar ground clutter. Neither group considered the possibility that the UFOs could have been secret aircraft engaged in a top-secret military exercise.

Other reports of flying triangles have come out of Great Britain which seems to have more than its share of these particular UFOs. The following is a sampling from a UK web site "December 4, 1996, Woverhampton: They saw another object that appeared to be approaching them from a northeasterly direction and traveling in a southwesterly direction. This object was described as matte black and shaped like a triangle with green lights around its perimeter. It was flying at a moderate speed. When it flew overhead it stopped and hovered. They observed a bright, white light on its underside that seemed to go out and then re-illuminate. Two men observed several green lights, one at each point and one in the center of each side. They thought the size was similar to a 747, but this craft made no noise at all.

November 4, 1996, Warrington, Cheshire: At 7:45 A.M. he observed a large jet flying south and banking to the southwest. Then he noticed a small black triangular object very close to the jet's tail, but moving independently. It would shift its position between the tail and starboard wing. At 7:25 P.M. another witness saw a black triangle moving fast. No sound was heard and no lights were visible.

October 8, 1996, Yetminster, Somerset: A man on the outskirts of Yetminster saw a "diamond shaped white light." The light was coming from a field to his left and seemed to follow him for about 400 yards. The next day, he called police and was told had been no helicopters in the area at the time.

October 4, 1996, Ramsgate, Kent: A woman was walking home when she saw a "very bright triangular object in the sky." It had bright lights at two points and a larger light at a third point. The object hovered for almost two hours. It had a greenish hue and appeared to revolve.

February 14, 1996, North Sea: A crew on a fishing boat saw a strange configuration of lights hovering above their vessel and moving slowly. About eight miles away, another fishing boat crew saw the lights and also detected an object on their radar system. The object was described as a black triangle with 15 red lights on each side and no lights underneath. It silently hovered for about an hour and then vanished.

One of the most famous triangle incidents occurred at the Manchester Airport on January 5, 1995. In UFO circles it became known as the "Manchester Near Miss." Two British Airways pilots were flying a Boeing 737. As they were coming in for a landing at the airport a dark, "wedge-shaped" aircraft passed them at high speed, startling them. The object was described as being lighted, dark, and with a black stripe down its side. It made no sound and the pilots felt no wake. The two pilots reportedly ducked as the object sped by their airplane on the right. Although the object passed very near to them, it made no effort to change its course.

There was no doubt the flying triangles were incredible objects seen by many credible people. But the objects were continually downplayed in official circles. For example, the. evidence in the Rendlesham Forest case was suppressed by the British MoD. Military officials had looked at the incident and judged it not to be of any military or defense significance. Despite headlines like "Sinister Plot to Hide The Truth" in British newspapers, the authenticity of Colonel Halt's audiocassette tape was considered to be suspect and the veracity of the witnesses was called into question. Once again, attention was diverted to the witnesses while the event itself was ignored.

Another famous UFO flap involving flying triangles took place in New York's Hudson Valley in the 1980s. The first sighting was reported on New Year's Eve 1982 in Putnam County and the last was on July 10, 1986 in southwestern Westchester County, Queens, and south central Long Island. The encounters are described in *Night Siege: The Hudson Valley UFO Sightings* by Dr. J. Allen Hynek, Philip J. Imbrogno, and Bob Pratt. In the Preface the authors state:

Many people believe it was a spaceship from another world. Whatever it was, it was described by many competent, professional people as a startling series of brilliant lights, generally in the form of a V or boomerang. It moved slowly and silently and was easily as big as a football field - some witnesses said as big as three football fields. That would make it anywhere from 300 to 900 feet long, far larger than any aircraft manufactured in the United States or any other country. This mysterious unidentified flying object ... was also very close to the ground, often no more than a few hundred feet in the air, and many people were able to get a good, long look at it.

On October 28, 1983 Jim Cooke, a biomedical engineer, saw some lights through the trees as he was driving home on the Croton Falls Road to Mahopac, New York at about 2: 15 A.M. The lights had approached very fast and they were very close to the Croton Falls Reservoir. As he slowed down to get a better look, he noticed the lights had apparently turned off, and all he could see was a "dark mass" behind the trees. "It looked like no aircraft that I ever saw before." The object was triangular in shape and was hovering about 15 feet above the water. "It had a highly non-reflective surface. I could see the dark mass, but I couldn't make out any detail at all." The object was about 200 feet away, but he could hear no sound.

Flying triangle photographed low over the landscape.

While Cooke watched, nine red lights lit up the sides of the triangle. Then a red beam of light, or else a solid object that was glowing red, came from the object's underside. He thought it appeared to be probing the water. The triangle moved to four different areas over the reservoir, and each time it shone a red probe on the water. The object stayed steady at about 15 feet above the water's surface. At times, especially when cars would go by, the lights would go out and the craft would become nearly invisible. Cooke thought the object was "at least one hundred feet long at the base and thirty feet at the apex." After it finished probing the water, it lifted up silently at a 33-degree angle until it disappeared.

The authors of *Night Siege* considered the possibility that the object "could have been a secret, experimental device originating from somewhere on Earth," but it's ability to "hover silently argues against that, unless some government had developed a technology that makes all aircraft, space shuttles, and booster rockets obsolete."

During the course of the UFO flap, some 5,000 people reported a triangular or boomerang-shaped object, as big as a football field, moving slowly or hovering, occasionally zipping away, and sometimes sending down a bright beam of light. The sightings covered a 1,400 square-mile area and included three counties in New York and three neighboring counties in Connecticut.

The first sighting was on New Year's Eve, 1982 when a retired New York City police officer whom the authors called Tony Vallor stepped outside of his house and saw a group of bright red, green, and white lights in the southern sky. At first he thought it was a jet in trouble, but then he realized the object was moving too slowly. He called to his wife to bring out the movie camera. When she came out, the lights were directly over their house. Vallor said he felt a "deep, vibrating sensation in his chest," and he could hear a "faint, deep hum."

As he began filming, the colored lights vanished and three bright white lights appeared in the shape of a triangle. "The lights were so bright 1 could see the ground all around me." Suddenly the white lights went out and the colored lights reappeared. Vallor counted fifteen of them. The object passed over them and slowly moved off. "1 never saw anything like this before, but 1 can tell you this was not any type of aircraft that I know of."

Another witness, Edwin Hansen, saw the same object just before midnight, a few moments after the Yallors. Hansen spotted some lights as he was driving home on Interstate 84. At first he thought it was a helicopter scanning the ground with a searchlight. He saw other cars had pulled over and were watching the craft. As Hansen watched, the object "began making slow, tight circles in the air," while "projecting a beam of light onto the ground." He realized it wasn't a helicopter because he couldn't hear any noise.

Hansen slowed down even more, peering through his windshield, trying to see if there was a shape to the lights. Then the object descended and headed straight for his car. "It was shaped like a boomerang with lights running up and down its wings. Part of what seemed to be a long, triangular tail section loomed behind the boomerang pattern of lights. It was so huge it filled up the entire sky." The craft seemed to be coming closer and he blew his horn at it. He tried to shield his eyes from the intense, bright light. Then another beam of white light shot down on the highway. By now, Hansen was truly frightened, yet he thought the craft seemed to be communicating with him, telling him not to be afraid. Then the object turned away and the beam of light disappeared.

On the night of February 26, 1983 Monique O'Driscoll, along with her seventeen-year old daughter, was driving home when the car radio began to hiss wildly. The static became so loud that she turned the radio off Then the two women noticed lights on a hill that at first they took to be a house. But the lights began to move slowly above the treetops. About 50 bright lights were moving slowly. They followed the lights to White Pond.

O'Driscoll pulled off to the side of the road and they watched the boomerang-shaped object stop right over the lake. "The lights were going like crazy, flashing. Red and blue lights were reflecting off the ice. There were a few amber lights in between, and the one in the middle was a big amber light."

Other witnesses appeared. The UFO began to move, swaying like a pendulum, then it started to move slowly away. O'Driscoll said the object was perhaps 200 to 300 feet "from tip to tip," and was bigger than her mother's house. She also got a look at its underside while she was beneath it. She described it as made of metal, diamond-shaped with "crisscross effects" and "tubular things here and there." Neither mother nor daughter heard any sound from the craft.

Shortly after that sighting, Herbert Proudfoot, a Danbury, Connecticut teacher, also saw the object. He was in a car with two friends who were air traffic controllers. Proudfoot was just getting off Interstate 84 when he noticed colored lights approaching them in the distance. His two friends also noticed the lights that formed a boomerang shape. He stopped the car and all three got out to watch the object. At one point it was directly overhead, about 1,000 feet above them, but they heard no sound.

"The lights seemed to be fixed to some type of solid structure, and the only motion it made was a slight wobbling at times. The underside of the object was dark black and looked like something out of the movie *Close Encounters of the Third Kind"* The authors wrote they would hear this repeated again and again. All three witnesses said they knew of no aircraft as large as this object, nor one that could hover the way it did. One of the air traffic controllers did think it might be an experimental government aircraft, but he thought it odd that it would be tested over a populated area.

Although the Danbury police knew about the sightings, they refused to cooperate with the authors of *Night Siege.* Said one officer, "Stable people do not report UFOs." However, the official attitude would change a year later when several Danbury police officers had their own close encounters. The authors would also discover that UFO sightings were part of Hudson Valley history. Sightings were reported in 1956, 1957, 1958, 1976, and 1978, and all reported boomerang-shaped objects.

During the flap, witnesses reported a large low-flying object with red, blue, green, and amber lights arranged in a triangular or boomerang pattern. Some also said a large bright light was in the center, underneath the UFO. Witnesses were also struck by the object's ability to hover. "I saw the thing, and it hovered directly over my head for five minutes -- and airplanes don't hover," said a Yorktown police officer. Also, the object made no sound, although a few witnesses described a "low engine sound" or a "low hum" that was barely audible. Some also claimed to have seen a brilliant beam of white light that streamed down from the craft and lit up the ground. Sometimes a small reddish object would come down along the beam and then shoot off.

Witnesses were amazed at the object's enormous size: "as large as a football field," as "large as an aircraft carrier," "larger than a 747," as large as three football fields, six stories high, and "If there is such a thing as a flying city, this was a flying city ... it was huge." All reported a triangular, diamond or delta shape. Several remarked on its ability "to hover silently and then zip away at unbelievable speeds, only to return in an instant." The UFO also seemed to have the ability to become invisible. Bill Hele, the chief meteorologist for the National Weather corporation, was on the Taconic Parkway when he saw the V-shaped object. He told the authors:

All the lights were changing color at a different time frame from the other lights, as if it had a rotating prism within the lights At this point all the lights on the horizontal axis ceased to exist

After not hearing a sound and seeing absolutely nothing in the sky except stars, I was amazed to find not even a silhouette of an object. That had me not exactly frightened, but on the verge. I'm searching the heavens all around me .. for thirty to forty seconds, and I 'm looking up when all of a sudden, flash! The entire string of lights came on.

A Greenburgh police officer, Mike Turnbull, and his wife saw the object while also driving along the Taconic Parkway. He saw the familiar boomerang pattern of lights and heard no sound. He slowed the car and noticed other motorists were pulling over and looking at the sky. Suddenly, the lights went out and he and his wife saw nothing. No structure, no visible object, "it was as if the object had become invisible." Then the lights just as suddenly came back on, but in a different part of the sky. The object increased its speed and flew off.

Cindy Tillison of Sandy Hook, Connecticut saw the object from her home. About 8:30 P.M. her Great Dane "started barking and went crazy." The dog was in the kitchen in front of the sliding glass doors, standing up and barking continuously, his ears straight up. When she looked out, she saw a "V-shaped object that was very, very bright. It was glowing an auburn color. It was huge."

Tillison watched it for about twenty minutes, getting more and more nervous, feeling like it was watching her. Suddenly a big beam of light shot down from the center of the craft. Not sure what to do, frightened, and unable to identify the object, she phoned her brother-in-law. But, just as she hung up, the object vanished. "I didn't see it go up. I didn't see it go down. It just disappeared, like nothing was ever there."

Gordon Gebert was driving south from Albany on Highway 87 when he saw a flash of light. Suddenly a large object was approaching his truck. "It startled me. It appeared out of nowhere ... it just drifted in toward me like it was a balloon of some sort, very graceful." He said the object was massive and boomerang-shaped with "white, red, and green lights all along the wing section." The UFO was flat black or a gun gray. Then, as suddenly as it appeared, it disappeared. "I lost sight of it for about a half second, and it was gone. This really blew my mind, like it vanished in thin air."

Two young women who saw the object said it made a faint humming noise as it went over them. They watched it zigzag across the sky until it shot straight up and disappeared. Said one, "When it disappeared, it went straight up real fast. All I saw was a streak of red and white light going up."

March 24, 1983 was the biggest night for the Hudson Valley UFO sightings. Thousands of people saw the object and police stations in the area were flooded with calls. For the most part, the police remained silent because silence had apparently become official policy. One year later, people again were reporting seeing the large, silent, triangular-shaped object. By that time, the UFO had been spotted in the area on and off for about fifteen months. The FAA said the lights were just stunt pilots and told people not to worry. Not only did people reject this lame explanation, but some were angered at the insult to their intelligence. Other people were concerned that no government agency seemed to be interested in the mysterious object. "If this thing can come in here and do whatever it wants, I want to know where the hell are our government's defenses," said Lieutenant Herbert Peterson of the New Castle Township police, one of three police officers who saw the V-shaped object

The sightings continued into 1984. In the summer of that year it was suggested that some government agency, quite possibly the CIA, was flying special aircraft in the area to cover up the UFO sightings. The CIA was believed to be using Stewart International Airport at Newburgh, New York for some secret operation. Residents, who lived near the base. reported seeing small black aircraft in hangers. One person even claimed he saw small black planes taking off in formation at night from the airport. But no one suggested that the UFO itself might be a top-secret aircraft. In fact, most believed that the object whatever it was, "was something beyond the technology of humans."

In September and October the UFO was spotted in the Brewster and Carmel area of New York. Some thought the government was aware of the UFO and conducting some type of secret operation. The U.S. government did own land in the area, and the National Security Agency used some of it. Perhaps the UFO was interested in the secret government work being done there. Area residents also spoke of "strange electrical surges, power outages, and all types of unusual TV interference that has been going on for many years," as well as "strange magnetic effects. on cars." And everyone was wondering where does a UFO the size of a football field go during the day?

Although triangle-shaped UFO reports were nothing new to ufologists, the Hudson Valley sightings were unique because they were reported by so many people in one area for such a long period of time, and all of the reports were remarkably similar and consistent. The sightings were widely reported in both the local and national media. Even the noted UFO skeptic Philip Klass wondered if this was an "unexplainable case." *The authors of Night Siege discounted aircraft, a secret weapon, a formation of planes, blimps or any man-made craft.*

"There is no conventional explanation for the Hudson Valley UFO ... we feel we have presented an overwhelming amount of evidence that something strange and inexplicable is in our midst; something alien to the world as we know it."

5. DEEP BLACK PROJECTS

I've sighted the thing! It looks metallic - and it's tremendous in size!
-- Captain Thomas Mantell, the first person to be killed chasing a
UFO

The hovering. the rapid accelerations, and the apparently effortless maneuvers of the UFOs clearly imply a far advanced technology. --J. Allen Hynek

       Kenneth Arnold didn't see flying saucers. He saw flying wings. The objects that Arnold described -- crescent-shaped, no tails, flying with an odd, weaving motion, no vapor trails -- all fit the flying wings. These strange aircraft, which had their roots in Nazi Germany, were just starting to be developed and tested in the U.S. during the immediate post war years. The Northrop YB-49 flying wing was crescent-shaped, had no tail, and looked like a straight line when seen edge-on. The YB-49's speed was only 495 mph, much slower than the 1,000 plus mph that Arnold had calculated for his flying disks. But there were likely other flying wing X-planes at the time that were capable of the speed Arnold had estimated.

       Flying wings were top secret aircraft, black jets that were kept hidden from the American people as well as from most government and military officials. The credible UFO reports that began on that June day in 1947 and continue to this day were and still are sightings of black aircraft; deep black projects that saw the research and development of highly advanced, highly unusual, and highly classified aircraft. The flying wings and the U-2 spy planes were the first of many projects including circular wing aircraft that were developed and tested under the strictest secrecy, with both the knowledge and participation of the CIA. The pilots who flew the first experimental deep black aircraft worked for The Agency, not the military. Donald Keyhoe had been right, but he had also been wrong. The CIA was involved, but it wasn't concealing information about UFOs that came from other worlds; it was concealing knowledge of UFOs that came from planet Earth.

       There is a direct evolutionary line from the first flying wings and the U-2 to the stealthy aircraft we know today like the F-117 and B-2. Unfortunately, many famous UFO incidents like Rendlesham

Forest, Belgium, and Hudson Valley occurred before these aircraft were made public, when aircraft like the F-117 and B-2 were still in the black world. One of the first top secret aircraft was the U-2 spy plane and we now know that many early UFO sightings were actually of this aircraft. The first operational stealthy aircraft was the top secret SR-71 Blackbird, then the most advanced aircraft of its time and still the highest and fastest aircraft ever made public.

From the beginning deep black aircraft were so highly advanced and unusual that they seemed to defy the very laws of physics. It was almost unbelievable that mere humans could make such futuristic machines. Flying much higher and faster than any known or conventional aircraft, they seemed to appear and disappear as if by magic. Because they flew at supersonic (hypersonic today) speeds, they often assumed a heat-streaked appearance, and would sometimes appear as glowing fireballs or flashes of bright light. They would show up unexpectedly on radarscopes and then mysteriously vanish. The pilots who flew these aircraft wore pressurized suits like the astronauts, giving them an odd, alien-like appearance. Test pilots were not only flying top secret X-planes, but were also involved in secret high altitude balloon experiments.

Witnesses who saw these aircraft, including experienced pilots, had no frame of reference to judge what they had seen. The aircraft were so exceptional in both appearance and performance that they seemed out of this world. They were so top secret that hardly anyone knew anything about them including the experts who were ostensibly investigating the UFO phenomenon like J. Allen Hynek and Edward Condon. Even top level military and government officials were unaware of black aircraft. And the officials and agencies that controlled black projects went to great lengths to suppress knowledge and evidence of them. Funding for these aircraft were purportedly hidden in budgets for legitimate aircraft. or space projects, thus making it easier to deny their existence as well as making it almost impossible to trace them. In addition, all kinds of ludicrous explanations from mountain mirages to weather balloons to temperature inversions to swamp gas were trotted out to appease an ever more suspicious public. Thus, when deep black aircraft were spotted, people were at a loss to comprehend, let alone, explain what they were seeing. They actually had no alternative but to call them what in fact they truly were -- Unidentified Flying Objects.

Kenneth Arnold's flying disks was one of the first sightings of flying wings. These sailplanes were the forerunners of stealth technology, and their history can be traced back to World War II. After conquering France, Adolph Hitler next turned his lethal attention to Great Britain, his last remaining obstacle to complete conquest of Western Europe. But before he could invade the island nation, he first had to destroy the Royal Air Force, something that turned out not to be so easy. Although Goering's *Luftwaffe,* sent over wave after wave of bombers, the RAF somehow always managed to give as good as they got. The quick and easy victory the Germans had anticipated and counted on kept eluding them. It took awhile for the Germans to realize that in addition to the bravery and skill of their pilots, the British had also developed a secret weapon. That weapon was a cutting edge, new technology called radar.

German fighter and bomber aircraft crossing the English Channel were picked up by radar and the RAF scrambled to meet them. The British knew the direction and number of enemy aircraft and this knowledge gave them a distinct advantage. If the Germans were going to win the war, they needed a new type of aircraft, one that could evade radar detection. Developing such an aircraft was given top priority and became part of Hitler's "wonder weapons" . Another segment of the wonder weapons program was the rocket project led by Vernher von Braun but while von Braun and his V-I and V-2 rockets would become well known, Reimar Horten and his flying wings would remain in the shadows. Yet Horten's flying wings were just as important as von Braun's rockets, perhaps more so.

Reimar Horten was born in Bonn, Germany in 1915. He studied mathematics and aerodynamics at the Technical Institutes in Bonn, Berlin, and Gottingen. In 1933 (the year Hitler came to power) Reimar and his brother Walter tested an all-wing sailplane (called a *nurflugel* or only wing). In 1937 the Horten brothers built a twin-engine machine called the Ho5. Unfortunately, it crashed on its first flight. The aircraft broke apart along with Reimar's jaw and one of Walter's teeth. The reasoning behind the *nurflugels* was that an airplane's fuselage and tail were unnecessary to lift so get rid of them.

But the problem with all-wing aircraft was that the center of gravity was so far back that the aircraft had a tendency to tumble over backwards, hence the crash. To overcome this problem, the brothers moved the engines forward while lengthening the propeller shafts. In 1938 the Ho5-B had a successful flight. With the advent of World War II, Reimar began exploring the idea of an all-wing transport glider for use in the invasion of Britain. But Walter had joined the *Luftwaffe* and he arranged a project in which Reimar would develop the *nurflugel* as a fighter-bomber.

In 1942, as part of a highly classified project, Reimar built the Ho9, an unpowered prototype with a 61 foot wingspan. The Ho9 was built at the Gotha factory in Friedrichsrode. After overcoming some difficulties, he used two Junkers turbojets like the ones that had been developed for the Messerschmitt Me262 (the world's first operational jet aircraft). In 1944 the jet-powered flying wing was flown successfully reaching a speed of 500 mph, almost as fast as the Me262. By comparison, the P-51 Mustang, the most widely used allied combat fighter in World War **II,** had a maximum speed of 445 mph and the long range bomber, the B-17 Flying Fortress, could only reach a top speed of 290 mph.

The Gotha Company readied the *nurjlugel* for as a fighter-bomber. It was to fly at 623 mph and would have made it the fastest aircraft of its day, even surpassing the Me262. This aircraft was designated the Ho229 and looked unlike any other known aircraft. It had no fuselage, nor tail and had a swept wing, batlike appearance. Its unusual shape, which lacked right angles, made it virtually invisible to enemy radar cross-section (RCS).

Reimar Horten was also working on another top secret project; the Ho 18 *Amerika bomber,* huge six engine batlike warplane that was to deliver an atomic bomb to Washington, D.C. or New York City. It would have been able to fly from Berlin to New York without refueling. Built in 1944, the Ho 18-A looked remarkably like a modem stealthy aircraft. Its wings were made of wood held together by a carbon tail which also helped make it invisible to radar. The Ho 18-B was another long-range bomber. Reimar Horten also built other *nurflugels.* Horten V was a jet powered version of a glider and the Horten IX was a jet fighter bomber with a speed of more than 600 mph and an altitude of 40,000 feet.

On April 14, 1945 the Gotha factory was captured by the U.S. Third Army, and U.S. Intelligence quickly moved in and sealed it off. Almost everything inside the aircraft factory including the Ho229 and several *nurflugels* were shipped to the US under tight security.

Walter Horten remained in Germany, but Reimar headed for Argentina and went to work for the Peron government. He became instrumental in Argentina's aeronautical industry and continued to build and test flying wings. Although Reimar Horten supposedly never left Argentina, he reportedly participated in the production of the B-47, the B-58, and the B-2. He died in 1993 in Argentina. In 1975 he was honored by Queen Elizabeth with the British Gold Medal for Aeronautics. Hence the suspicion he had been involved in the UK's stealth program.

In the U.S. Jack Northrop had been experimenting with flying wings since the end of the 1920s. In 1940, he flew the twin- engine N-1M Flying Wing. In 1942 the N-9M appeared. In the Soviet Union there were also attempts to develop all-wing aircraft. Boris Ivanovich Chemanovski who produced a series of projects from 1921 to 1940 was the most successful Soviet designer. . Despite the fact that design, research, and development on flying wings were going on in three separate countries at the same time, there was no collaboration at all.

In 1939 Jack Northrop founded the Northrop Corporation where he served as President, Head of Engineering, and Chief of Research. His first flying wing was followed by the giant B-35 that had a wingspan of 172 feet. Its first flight was in June 1946. He also developed the X-56, a welded magnesium fighter and the Black Widow, the first American night interceptor. He died in 1981.

Glen Edwards was an Army Air Force test pilot who had been a 50-mission veteran of World War II. He was one of the test pilots who qualified to fly the N-9M, the much smaller version of the B-35. Edwards reported to the Air Technical Service Command that the N-9M "comes off the ground of its own accord between 70 and 75 mph and immediately assumes a steep nose high altitude." In his diary he wrote that flying the "little beauty" was "quite an experience, quite different from flying anything else." .

The wings were tested at California's Murdoc Army Air Base, a dry lake bed in the high, isolated desert country. Edwards was given command of the YB-49 flying wing program. On June 5, 1946 Danny Forbes and Glen Edwards took off in their flying wing with Forbes as aircraft commander and Edwards as copilot. The YB-49 went into a tail slide and the wing tumbled over backwards or sideways into a high-speed, maybe even supersonic spin. Its outer wings tore off. The two test pilots crashed into the desert northwest of Murdoc and were killed. To honor Glen Edwards, Murdoc was renamed Edwards Air Force Base and became the home of the Air Force's Flight Test Center where many deep black projects were flown and tested.

On June 26, 1947 the *St. Louis Post Dispatch* carried the following article: *Pilot Reports Seeing Weird Objects in Sky: Army and Civil Aeronautics Administration spokesman expressed skepticism today over a report of nine mysterious objects - as big as air planes whizzing over western Washington at 1,200 miles an hour. But Kenneth Arnold, a flying Boise Idaho, businessman who said he timed them by his plane clock, clung to his story of the shiny, flat' objects, each as big as a DC-4 passenger plane, racing over the Cascade Mountains with a peculiar weaving motion like the tail of a kite. An Army spokesman commented, As far as we know, nothing flies that fast except a V-2 rocket, which travels too fast to be seen.*

Kenneth Arnold had unwittingly become the first American to report seeing deep black aircraft. And like the many witnesses who would follow him over the years and decades to come, he had no idea what he had seen. No military or government official came forward to clear up the confusion despite the fact that flying wings were not entirely unknown to the public. The 1953 movie, *War of the Worlds,* featured a flying wing. The aircraft carried the bomb that was to be dropped on the invading Martians. There is some cool stock footage of the wing in the film's final half-hour. Also, a Paramount newsreel shown in 1949 reported on a Northrop Flying Wing, the most powerful and longest range and fastest jet aircraft in the world. The newsreel predicted it would become the passenger plane of the future. Then, apparently, flying wings disappeared from public view and no more was heard of them.

Circular-wing aircraft were another part of Nazi Germany's

Wonder Weapons program. This project was under the leadership of Dr. Walther (or Richard) Miethe who worked at two facilities near Prague. In May 1945 a prototype was tested. Like the flying wings, circular-wings were being developed for their "stealth" capability. It was believed that the circular shape could reduce RCS even more than the triangular, flying wing shape.

Dr. Miethe probably worked with Dr. Werhner von Braun on his rocket projects before he became director of the circular wing program. The two may have even been good friends. Both men were purportedly on the Allied "Black List" of top German scientists whom the Allies wanted to get their hands on. "Operation Overcast" began in May 1945 with teams of Allied intelligence officers fanning out across Germany targeting top scientific personnel. Overcast had a threefold purpose: to enlist these scientists and technicians in the war against Japan, to keep them out of the hands of the Soviets, and to keep them from rearming Germany. In March 1946, Overcast was changed to Operation Paperclip because the first code word had been compromised. Paperclip enabled government officials to grant full immigrant status to the scientists and to allow them to bring their families to the U.S. By May 1948, there were some 1,136 German and Austrian nationals in the U.S., 492 were scientists and technical experts and 644 were their dependents.

Paperclip scientists were mainly involved in the rocket programs, but other projects also utilized these people. Walter Dornberger who had been von Braun's superior worked for the USAF and then joined Bell Aircraft Company. He worked on the Bell X-I and the "Dyna-Soar" hypersonic aircraft, and then went on to become Vice President of Bell Aerospace Corporation. Dr. Miethe also reportedly worked in the U.S. and was quite possibly involved with saucer-type aircraft including Avrocar, a joint U.S. and Canadian craft that reportedly failed because of a stability problem. However, other saucer programs followed Avrocar like the Silver Bug program that may have been based on Dr. Miethe's research.

In 1954 CIA Director Allen Dulles appointed Richard M. Bissell to coordinate Project Aquatone, the code name for the U-2 spy plane program. By the mid-fifties the Cold War was entrenched. The countries of Eastern Europe had communist dictatorships and were also occupied by the Red Army. The United States and its allies had been to the brink of war with the Soviet Union over the Berlin blockade. The Russians now had both the atomic and hydrogen bombs and the long-range bombers capable of delivering them to America's heartland. The Soviet Union was also a highly secretive and closed society. Even the Moscow telephone directory was classified. A climate of fear, suspicion, mistrust, and paranoia had grown between the two superpowers that had so recently defeated Nazi Germany together.

President Eisenhower was deeply embroiled in the Cold War, but he was also mindful of the fact that a costly weapons program could destroy the American economy. He needed to know just what weaponry the Soviet Union had and what it was up to. He didn't want to rely upon speculation and rumor. Eisenhower needed to see behind the Iron Curtain. Could reliable, objective information be obtained by flying over the totalitarian Soviet Union with an unarmed civilian spy plane?

The idea of conducting secret, high altitude reconnaissance flights over the Soviet Union was not new; Nazi Germany had employed the method first. In October 1940 Lieutenant -Colonel Rowehl was given a top-secret order from Hitler: "You will organize long-range reconnaissance formations, capable of photographic reconnaissance of Western Russian territory from a great height. This height must be so exceptional that the Soviets will not notice anything. You must be ready by 15th June I 1941." (8) Germany invaded the Soviet Union on June 21, 1941..

In the winter of 1940-41 the *Rowehl Geschwader* began its secret overflights. He-III aircraft with specially designed high altitude engines were used. They were also equipped with pressurized cabins and special cameras with wide-angle lenses. Another top secret aircraft was the Dornier Company's Do-215-B2 with a ceiling of 30,000 feet. Special Junkers models, the Ju-88B and Ju-86P, had ceilings of 33,000 and 39,000 feet, tremendous altitudes for that time. Large swaths of the western Soviet Union were photographed with these first spy planes without the Russians ever noticing. All the Soviet airfields including well-camouflaged fighter bases, army units in forward positions, and large concentrations of tanks and armor were photographed. These photographs played a decisive role in the opening of the Russian campaign; the Germans were able to destroy Soviet defenses at the outset.

After the war, the Americans retrieved the documents that detailed Germany's secret aerial reconnaissance flights. The documents, coded "Reconnaissance under the C-in-C Luftwaffe," were quickly buried in U.S. secret archives.

As the Cold War intensified, the U.S. also began to monitor the Soviet Union through the use of reconnaissance balloons. Project Mogul was a highly classified spy balloon used for monitoring the atmosphere over the Soviet Union for nuclear tests. It was the crash of one of these balloons in New Mexico in 1947 that caused the now legendary Roswell UFO Incident. The Soviets were well aware of these spy balloons and had developed a high altitude aircraft, the M-17 "Mystic," for shooting them down. The United States had also been using conventional bombers to spy on the Soviet Union. Although only cameras were in these aircraft, to the Soviets they looked like a threat. Eisenhower feared these spying methods might start a war. What was needed was a new spy plane, one that could fly so high and fast it couldn't be detected.

The man recruited by Eisenhower, the CIA, and the air force for this task was Clarence Leonard "Kelly" Johnson, the head of Advanced Development Projects for the Lockheed Corporation known as the Skunk Works.

Kelly Johnson was born on February 27, 1910 in Ishpeming, an iron-mining town in the Upper Peninsula of Michigan. He earned a B.S. (1932) and M.S. (1933) in aeronautical engineering at the University of Michigan and then joined Lockheed in 1933. At Lockheed he worked as a tool designer, flight test engineer, stress analyst, aerodynamicist, weight engineer, and wind tunnel engineer. In 1938 he became Chief Research Engineer. He would eventually head the Skunk Works, Lockheed's super secret aviation development department where many deep black aircraft were born.

Johnson had already designed the F- 104, the "missile with a man in it." The F -104 had tiny, sharp-edged wings with razor thing edges so that it could fly mach 2 (about 1,300 mph). In order to build the new spy plane, Johnson said, "I'd take the F-104 and give it wings like a tent." Eisenhower decided it would be best if the CIA handled the project. Code-named Aquatone, it would be one of the most secret projects ever undertaken by the United States government. Johnson wanted a plane that could fly higher than 70,000 feet and farther than 3,000 miles. Johnson's team set up shop in the top secret Skunk Works.

Johnson's plane had a long nose and narrow fuselage. The total wingspan was 80 feet. The fuselage was built of a very thin aluminum and it was 50 feet long. He wanted an extremely light weight plane; every pound less the U-2 weighed, another foot higher it could fly. It was essential that the U-2 be able to fly as high and fast as possible for its overflights would be violating Soviet air space.

Test pilot Tony LeVier was given the job of finding a suitable test site for the new aircraft. He chose a secret site in Nevada that Johnson called "Paradise Ranch." Today it is known as Groom Lake or Area 51. On August I, 1955 the U-2 Number One was rolled out for its first test flight with Tony LeVier as the pilot. LeVier took the plane up to 70 knots, chopped the power, pulled back the throttle, and was suddenly in the air. The U-2 seemed to just take off on its own. LeVier later told Johnson, "I took it up to seventy, and I didn't realize it at all, and this damn thing took off and I absolutely did not know it." The plane's take-off seemed to be over before it began, and the aircraft appeared to climb as if it was standing on its tail.

Johnson had promised the CIA he would have its plane in the air in eight months. It was time to build a fleet of U-2s and recruit the pilots who would fly the actual missions over the Soviet Union. Twenty-nine pilots arrived at Paradise Ranch. Their mission was so secret that they hadn't even been told what it was. Disguising their military backgrounds, they had resigned from the USAF to work for the CIA. The agency called this "sheep-dipping." One of the pilots was Francis Gary Powers. These men and their aircraft became the CIA's secret air force undertaking missions that remain classified to this day.

Because the U-2 operated on the fringes of the atmosphere, the pilots wore pressurized suits and needed supplemental oxygen. They had to observe radio silence while navigating on their own for up to ten hours. Every day they set altitude records. Said pilot Bob Ericson, "The view from up there -- the sky turns a dark, dark blue. You look straight down, and it would just be like normal daylight, but you looked at the horizon or up at the sky and a lot of times you could even pick stars out. And you looked straight ahead, you can see the curvature of the earth. "

The U-2 camera took pictures of the Earth from thirteen miles high. "It could bring into focus, objects as small as two and one half feet across, and take a series of photographs along a flight path three thousand miles long." The pilots practiced their top-secret missions by flying across the 'U.S. using a panoramic camera that photographed in seven different positions. Each mission carried a strip of film that was a mile long. As they flew high overhead, they photographed wide swathes of the country. While doing this, they generated sightings and reports of UFOs. In 1997 the CIA admitted that Blue Book investigators knew many UFO reports were actually sightings of the U-2, but couldn't make the knowledge public .

The first mission over the Soviet Union came from a USAF base in Wiesbaden, West Germany. Few of the base personnel even knew the secret plane was there. On July 4, 1956 the U-2 flew over the Soviet Union for the first time, in violation of Soviet air space, The pilot was Harvey Stockman who was well aware that "This is another guy's air." At Minsk, Stockman turned north, heading toward Leningrad. Suddenly he spotted MIGs that had been sent to attack him. He had been spotted on Soviet radar. Kelly Johnson had assured the pilots that the Russians were not going to be able to touch them for at least two years. And Johnson was right. The U-2 flew too high for the Soviet fighters to catch, but the spy plane was no longer a secret. The Soviets protested secretly to U.S. government.

Back at the Skunk Works, Kelly Johnson and his team had to cope with the unhappy fact that Russian radar had picked up the U-2 on its very first mission. The plane's large flat wings were easily picked up in radar cross section so Johnson's team embarked on the "Dirty Bird Project." They covered the spy plane's belly with a metallic grid and special paint to absorb radar waves. In April 1957 pilot Bob Sieker flew a Dirty Bird at 72,000 feet, but the heat built up rapidly and then the engine quit. Sieker had to bail out, but because there were no ejection seats back then, he had to climb out of the aircraft. He was hit by the tail and killed. It took Johnson's team three days to find the wreckage. Sieker was the first of ten fatal accidents that would take place during flight testing and training. But the missions over the Soviet Union were successful, yielding a wealth of military intelligence about the country.

Then in October 1957 the Russians launched *Sputnik* and the Americans panicked. Eisenhower now wanted the new technology of spy satellites, which could cover more territory than a U-2. Also, there were reports indicating the Soviets now had SAMs (surface-to-air missiles) that might be able to climb high enough to reach the U-2s.

The CIA wanted a new U-2 mission that would overfly a long-range missile base in the northern Soviet Union. Eisenhower approved the mission that would be the longest U-2 mission ever. The pilot chosen was Francis Gary Powers. Taking off from Pakistan, Powers was to fly over the *Sputnik* launch site in Tyura Tam, the defense installations in Sverdlovsk, and then the long range missile site. On May I, 1960 Powers left on his mission.

In Moscow, Nikita Khrushchev learned that once again Soviet air space had been invaded by a U-2. The premier was furious, especially because the spy plane had been sent on its mission during a Soviet national holiday (May Day). Powers was shot down over Sverdlovsk. Eisenhower approved the cover story which was that the plane had strayed off course while conducting weather research. But hopes for a test-ban treaty and for detente between the two countries were destroyed.

Powers was tried for espionage and sentenced to ten years in prison. He was released after seventeen months in exchange for a Russian spy. On the same day that Powers was sentenced, the CIA recovered film from the first of its successful spy satellites. The agency had discovered that one satellite mission had covered more Soviet territory than all of the U-2 missions together. It looked as if the CIA might be getting out of the spy plane business.

But one year before Francis Gary Powers had been shot down, Kelly Johnson and Lockheed already had been named to develop the follow-on aircraft to the U-2. This project, code named Oxcart, saw the beginning of the stealth technology that would revolutionize aircraft design and would be responsible for most of the credible UFO sightings from the 1960s to the present day.

The flying wings were about to come into their own.

6. BLACKBIRD

Okay, go sleep it off and the pink UFO will go away. Danbury, Connecticut police officer responding to a caller reporting a UFO, in *Night Siege: The Hudson Valley UFO Sightings* by Dr. J. Allen Hynek and Philip Imbrogno

On April 17, 1966 two sheriff's deputies, Dale Spaur and Wilbur Neff were finishing an accident investigation near Ravenna, a town in eastern Ohio. At 4:50 A.M. they were told by the Portage County dispatcher to be on the lookout for a low-flying UFO heading in their direction. Driving west, they came upon an abandoned car and stopped to investigate. That's when Spaur, a Korean War veteran, spotted a glowing object about 1,000 feet above the trees. Both deputies watched as the object grew larger and moved south. Then it changed direction and came right toward them, lighting up the entire roadside. Neither man had ever seen anything so bright. They ran back to their squad car and Spaur contacted the dispatcher to report what they were seeing:

"It's about fifty feet across, and I can just make out a dome or something on the top, but it's very dark. The bottom is real bright; it's putting out a beam of light that makes a big spot underneath. It's like sitting on the beam. It was overhead a minute ago, and it was bright as day here. And this is no helicopter or anything like that. It's perfectly still and it just makes a humming noise."

The dispatcher told the deputies to keep the UFO in sight. When the object moved, they followed in their patrol car and soon the two men were racing along at almost ninety miles an hour. They headed east toward Pennsylvania and were soon joined by Wayne Huston, a police officer from East Palestine, Ohio. Huston had been listening to their radio traffic and he waited for them at an intersection. They saw the UFO fly over, speeding along at over eighty miles an hour at an altitude of about 900 feet. The object crossed the state line. Huston notified the Pennsylvania State Police who contacted the Greater Pittsburgh Airport to see if the object was on their radar, but the air traffic controllers said there was no UFO on their screens.

Spaur and Neff were now 85 miles from where their chase had begun and they were running low on gas. They pulled into a service station near Conway, Pennsylvania with Huston behind them. Patrolman Frank Panzanella of the Conway Police Department who had also been chasing the UFO joined them. Standing together, the four police officers watched the UFO hovering in the east. The object moved higher and they saw a commercial aircraft pass beneath it. Pananella called the Pittsburgh airport control tower and this time the radar operators said they had picked up the object. Then the UFO suddenly shot upward and disappeared. It was later reported that another policeman had seen two jet fighters being followed by "a bright object shaped like a football."

Before Spaur and Neff returned to Ohio, they called on an air force reserve officer at the Pittsburgh airport who briefly interviewed them and then told them a report would be filed with Project Blue Book. When they got back to Ravenna reporters were waiting for them. The wire services had been monitoring the police radio transmissions, and the UFO chase had caused a sensation. Spaur was willing to talk to the media, but Neff went home and wouldn't talk to anyone.

The next day Major Hector Quintanilla, head of Project Blue Book, called Spaur and asked him about the mirage he had seen. Five days later, the air force announced that the officers had been pursuing a communications satellite called Echo and then the planet Venus. The Pittsburgh Airport also denied that it had picked up the UFO on its radar.

The public outcry over these inept explanations was so great that Major Quintanilla went to Ravenna and personally spoke with deputy Spaur. But, he made no effort to talk to the other involved in the chase, and he soon left. Police Chief Gerald Buchert of Mantua, Ohio claimed he took a picture of the UFO, but the air force told him not to make the photo public. Quintanilla held the position that UFOs were misinterpretations of conventional objects and natural phenomena, especially astronomical phenomena. The police officers had simply been mistaken.

Outraged by this offhand attitude, the officers felt their credibility had been called into question and they were being made to look ridiculous. Spaur soon resigned from the sheriffs department and got divorced. Commented Spaur: "If I could change all that I have done in my life, I would change just one thing, and that would be the night we chased that damn saucer. " Neff would have nothing more to do with the incident. "If that thing landed in my backyard," he told his wife, "I wouldn't tell a soul."

Another officer involved in the chase resigned from the police force and moved to Seattle. "Sure I quit the force because of that thing," he said. "People laughed at me, and there was pressure. You couldn't put your finger on it, but the pressure was there."

J. Allen Hynek was also dissatisfied with Quintanilla and Blue Book's handling of the case. Hynek felt it was a classic example of a Close Encounter of the First Kind. Unfortunately, it was also a classic example of how Blue Book handled genuinely mystifying cases -- dismissing them and treating them with little or no interest while thoroughly investigating and reporting cases that were obviously misidentifications, misinterpretations, or outright hoaxes. In this instance, the case was listed in Blue Book as an observation of Venus.

On August 25, 1966 the officer in charge of a North Dakota missile crew was based in a concrete capsule 60 feet underground. Suddenly he found his radio transmission interrupted by static. At the same time, other air force personnel on the surface were reporting a UFO sighting. The object was described as a bright red light alternately ascending and descending in the sky. Members of an air force radar installation on the surface also reported the and tracked it at an altitude of 100,000 feet. Then the UFO appeared to swoop and dive and seemed to land ten to fifteen miles south of the missile site. A strike team of well-armed air force guards was dispatched to check the area where the object had landed. When the team was about ten miles from the area, the object's glow decreased and the UFO abruptly took off. Another UFO was seen and also confirmed by radar. The first object that had been sighted flew beneath the second one. This maneuver was also confirmed by radar. The first one then flew toward the north while the second one disappeared in a glow of red. This incident was investigated by Dr. J. Hynek who called it "typical of the puzzling cases" he has studied.

In the winter of 1966 a DC-8 commercial airliner was on a regular flight from Lima to Mexico City. Four crew members reported an odd sighting to the left of the airplane. The captain reported:

Two very bright lights, one of which was pulsating; from the two lights were two thin beams of light which moved from a V initially to an inverted V finally. At one point the object seemed to emit a shower of sparks. There appeared to be a solid shaped object between the two white lights, which was thicker in the middle and tapered outwards. There was also a strip of light between the white lights, not very bright and yellowish in color. Much like the cabin lights of an aircraft.

Investigator Franklin E. Roach of the Condon Committee wrote: "The apparent pacing of the aircraft by the object for an estimated two minutes is a puzzling feature of the sighting. Also the captain's sketch is suggestive of some kind of a craft. These added up to the intriguing possibility of an intelligently guided craft which, in the words of the aircraft's captain: "is a craft with speed and maneuverability unknown to us." The captain believed the object was definitely not an aircraft. Roach concluded that the most likely explanation was that the sighting was the "result of the reentry of fragments of the Agena from Gemini 11."

For two weeks each in 1984 and 1985, Scandinavian researchers conducted a two-part study on UFOs that had been seen over the Hessdalen Valley in Norway, five miles below the Arctic Circle. Beginning in December 1981, villagers .reported seeing several UFOs in the sky. During one five week period in January and February 1984, as many as 188 sightings were reported. Villagers saw strange lights, ovals, and cigar-shaped objects. The Hessdalen crew picked up several of the UFOs on radar, sometimes even when the objects were not visually observed. They also took long-distance photographs.

During this time a Canadian psychophysiologist, Michael A. Persinger, proposed a theory that geophysical processes associated with faults in the earth's crust created "earth lights" which some people were interpreting as lights from spacecraft. This tectonic activity, the underground movement of the earth along fault lines, compressed quartz crystals in rocks that released a form of energy known as piezoelectricity. According to Persinger, this energy could produce balls of light capable of lasting duration and unpredictable behavior. Hessdalen is located on a fault zone.

On September 3,1959 Lockheed's Skunk Works with Kelly Johnson in charge was given the go-ahead to develop a follow-up aircraft to the U-2 spy plane. The original design was designated the A-12 and was to be a high-speed fighter interceptor. It won a highly secret design competition. But the CIA wanted a reconnaissance aircraft, not a fighter. Johnson modified his design and the CIA approved it. The project was code-named Oxcart. The agency wanted an aircraft with extremely small RCS (radar cross-section) so that it would be virtually invisible to radar. In order to accomplish this, Johnson designed an aircraft capable of fast speed and high altitude and made with radar absorbing material or RAM. The aircraft would be built with titanium.

As the aircraft was being designed, the test site for the plane was being modified on Groom Dry Lake, the Nevada site first chosen by Tony LeVier to test the U-2. Because the CIA funded the project and because the aircraft was so technologically advanced, the program was highly classified. It was so top secret that even the Joint Chiefs of Staff purportedly knew nothing about it.

Like the U-2 program, the test pilots chosen for this project were also top secret. The pilots did not know what they would be flying. The ten pilots chosen were sent to Worchester, Massachusetts where they were outfitted with personal S-90 I full-pressure suits. These pressurized suits were just like those worn by the Mercury and Gemini astronauts. On April 26, 1962 the first test flight of a high altitude Mach 3 aircraft was made at Groom Lake. It lasted 35 minutes. This was the A-12, and the YF-12A program quickly followed it.

Although YF-12A never became operational, it set a speed

record of 2,070.101 mph and an altitude record of 80,257.86 feet. Like the A-12 the YF-12 A was the forerunner of the SR-71 Blackbird. To withstand skin temperatures in excess of 500 degrees Fahrenheit created by air friction, most of the YF-12A was made of titanium alloys, and it was coated with special black paint that helped to radiate heat from the skin.

The SR-71 Blackbird's airframe was also made almost entirely of titanium alloys. Flying at speeds in the Mach 3-4 range with a ceiling of 85,000 to 100,000 feet, it operated at the edge of the Earth's atmosphere. Its range was between 3,000 and 4,000 miles. Because it flew at such high speeds, the cockpit windshield could reach a temperature of 326 degrees Fahrenheit and the wing areas could reach 426 degrees, hot enough to melt lead. Some of the structural components reached temperatures as high as 5,432 degrees. Electric switches and wires were gold-plated in order to increase the conductivity at such high temperatures. Tires were filled with nitrogen and impregnated with powdered aluminum to prevent them from exploding from heat buildup during flight. The aircraft was painted black to help lower temperature. Radar absorbing materials were applied to the leading edges of its wings, and the fuselage was made to slope outwards. The length was 107ft.5in., the wing span was 55ft.7in., and the height was 18ft.6in.

The needle-nose contained reconnaissance sensors, and the entire unit could be detached for different sensor combinations. Separate cockpits housed the pilot and the reconnaissance systems officer (the RSO). Although a computer largely controlled the sensors, the RSO had to monitor the operation. Four compartments in the fuselage housed panoramic, long-range, and infrared cameras, electronic intelligence sensors and side-looking radars. The aircraft could look down from above or peer sideways hundreds of miles into enemy territory. It could survey more than 100,000 square miles in an hour.

The SR-71-Blackbird-had.huge.J58 engines which operated as ordinary jets at low speeds, but then would switch to become ramjets at speeds above 2,000 mph. To witnesses, it appeared as if the aircraft was flying slowly, even hovering, only to take off at a tremendous speed and then disappear. The aircraft's black paint was specially formulated to radiate excess heat while at the same time disrupting incoming RCS. When it flew high and fast, its black changed to blue, probably accounting for those reports of blue-streaked UFOs.

After takeoff, the Blackbird would climb subsonically to rendezvous with a specifically modified KC-135Q tanker about seven minutes later. Usually two T-38s monitored the takeoff from above and behind. After filling up, the Blackbird would dive briefly and then accelerate to go supersonic. It then climbed to about.82,000 feet where it would cruise at Mach 3. It would then either refuel or cruise to the reconnaissance target. When it approached its target, the reconnaissance sensors were activated automatically by onboard computers that were tied in with the inertial and stellar navigation systems. It flew so high that it getting within range of enemy defenses. No Blackbird was ever lost on a mission although it was reportedly shot at over one thousand times. Most of the missions are still classified. After completing its mission, the Blackbird would descend to another tanker for more fuel.

On long-range missions, the aircraft might need refueling or four times. It was these practice mid-air refueling that were most likely responsible for the strange light configurations people were seeing in the night skies and reporting as UFOs. At that time air-to-air refueling could look bizarre--one plane would appear to be standing on its tail. Today tanker aircraft still performs air-to-air refueling. The Blackbird's high speeds also gave it a heat-streaked or fireball appearance. High speeds and high altitude would also change its color to blue. Add these to the fact that the Blackbird would accelerate and dive and accelerate again only to suddenly disappear as if by magic and you have a classic UFO sighting. Diamond shaped patterns appeared behind the engine due to the 60,000 pounds of thrust. This pattern would be visible during the daylight and could stretch 20 or more feet from the tail at night.

When the Blackbird landed the pilot had to trigger the drag chute to slow the aircraft down to taxi speed. Then the aircraft was taken to a special hangar where it was parked. Cooling fans were aimed at it to cool down the aircraft's still super-hot skin. As it would cool and contract, the skin popped and crackled. If the Blackbird had to land where special hangars were unavailable, the crew would remain inside with their suits' air conditioning systems running until the aircraft was cool enough for the recovery crew to help them out.

No wonder people who spotted the Blackbird or other black aircraft thought they were seeing alien spaceships. In May 1961' George W. Earley, a representative of NICAP (the National Investigations Committee on Aerial Phenomena), said in an address to the American Society of Mechanical Engineers:'

After all, the speed and maneuverability displayed by these UFOs calls for propulsion systems far in advance of anything we now have. The entire vehicle represents, in terms of present earthly knowledge, a tremendous technological break-through. Such a break-through would be quickly reflected in hundreds of allied fields, as well as in fields never dreamed of before ... So we are left with the interplanetary theory.

In building the SR-71 Blackbird, Kelly Johnson and his team had created the first stealth aircraft, developing a technology to reduce the RCS of a reconnaissance or combat aircraft. This fast, high-flying aircraft could take-off from one of its three main bases and be at any point on the earth's surface in just six hours. By one estimation, it cost $200,000 an hour to fly the Blackbird. The total operating cost was estimated between $200 million and $400 million a year. It is believed that a total of 32 aircraft were built. By 1988 only six Blackbirds remained operational and the aircraft was withdrawn from service in 1989, although three planes are believed to be still used by NASA for research and possibly by the air force for reconnaissance. In 1990 a Blackbird crossed the United States at a record speed of 2,124 mph. Allocated to the Smithsonian Institute, it flew from Los Angeles to Washington, D.C. in 64 minutes and 5 seconds.

Like the U-2 spy plane before it, the Blackbird also operated out of the UK, flying regularly out of RAF Mildenhall in East Anglia where it used a 9,000-foot runway. The UK was about the only NATO country that allowed the top-secret aircraft to operate from its soil and in its airspace. Its deployment from the UK began in 1976 and lasted until 1980. The willingness of the UK to let the Blackbird operate from its bases illustrates the "special relationship" between the two countries.

Lockheed's Blackbird SR 71

The Blackbird was preceded by a high-altitude, long-range aircraft known as the CL-400 Suntan developed during the late 1950s. Also built at the Skunk Works, the Suntan could fly at Mach 2.5 with a range of 2,500 miles, and had a revolutionary hydrogen propulsion system. Although the air force spent as much as $1.2 billion on the top-secret program, it was canceled in February 1959, probably because the aircraft could not meet a range requirement of over 3,300 miles. Nonetheless, it is highly likely that the Suntan and/or its prototypes, along with the U-2, were responsible for many of the credible UFO sightings in the mid to late 1950s, especially those reported by pilots.

One version of the SR-71 Blackbird carried the 0-21 reconnaissance drone. The drone was an unmanned aerial vehicle (UA V) that was launched in flight at supersonic speeds. While black aircraft were kept top secret, UAVs were kept even more top secret. UAVs launched from Blackbirds and quite possibly other black aircraft were no doubt responsible for their share of UFO reports, especially those cases where witnesses described a smaller light or object falling from a larger light or object.

In 1989 the Skunk Works was developing a reconnaissance UAV. It was a flying wing with a span of 85 feet. This program code-named "Q" was top secret and funded by Lockheed. After Desert

Storm, the air force funded a "Q" demonstrator as the Advanced Airborne Reconnaissance System (AARS). The cost was expected to be about $150 million per vehicle.

In 1994 Boeing teamed with Lockheed Martin to design and build DarkStar, a stealthy, unmanned reconnaissance aircraft for the U.S. Department of Defense as part of the Tier III Minus program. By the time of DarkStar, Boeing already had three decades of UAV work and experience behind it. Boeing built the wings, provided its avionics, and integrated its radar and vehicle management system software. The Skunk Works built the fuselage and integrated the Electro-optical sensor and a number of subsystems.

On March 29, 1996 DarkStar, which looks like a flying saucer with wings, made its first flight, reaching an altitude of 5,000 feet. A fully automated flight was successfully made from takeoff to landing using the Global Positioning System (GPS). DarkStar is fifteen feet long and spans 69 feet. Its range is more than 500 nautical miles and its ceiling is more than 45,000 feet. It can also stay on station for more than eight hours, and is used for reconnaissance in highly defended areas. In January 1999 the Department of Defense announced that DarkStar was terminated. More advanced surveillance drones like the Predator and Global Hawk replaced it.

Another unmanned black project was Brilliant Buzzard, built secretly by Boeing in the 1980s. It was probably developed at Boeing's Phantom Works, which is similar to Lockheed's Skunk Works. Brilliant Buzzard was over 200 feet long and could operate unmanned at a cruising speed of Mach 3. It was the platform for the Airborne Optical System (AOS), an essential component of the Strategic Defense Initiative (SDI) missile defense system. (Anything with "Brilliant" in its name indicates SDI). It was top secret and rarely discussed because AOS violated the ABM treaty, prohibiting mobile components of a missile defense system. AOS was to pick up incoming warheads and determine their track, then cue the ground-based radar (GBR), separate the real warheads from the decoys, and direct the attack. The technology for Brilliant Buzzard had been ongoing for some 30 years.

It is impossible to say exactly what the two police officers from Ohio saw in 1966 when they were chasing that "damn UFO." But that was the same year the YF-12A test program ended and the SR-71 Blackbird became operational. The A-12 first flew in April 1962 and shortly after its test flights began, sure enough, witnesses including airline pilots began reporting sightings of UFOs. Both the U-2 and SR-71 Blackbird flew high altitude spy missions for years before they were officially revealed, and then only because even more top secret aircraft had replaced them.

The police officers didn't know about stealth. Very few people did. The stealth project was even more top secret than the Manhattan Project that had produced the world's first atomic bomb .. Deeply concealed in the black world, stealth aircraft would not enter the white world until shortly before the Gulf War, but when it did, it would reveal one of the greatest technological advances of the 20th century, and it would play a significant role in ending the Cold War.

7. STEALTH

I do not believe, let me stress, that this has been part of some top secret cover-up of extensive investigations by Air Force or security agencies. I have found no substantial basis for accepting that theory of why the Air Force has so long failed to respond appropriately to the many significant and scientifically intriguing UFO reports coming from within its own ranks.
__ Dr. James E. McDonald addressing the 1969 meeting of the American Association for the Advancement of Science.

There was a reason why President Bill Clinton and British Prime Minister Tony Blair believed that air power alone could win the war in Kosovo. The reason was stealth technology. It was during the Kosovo air campaign in 1999 that the B-2 Spirit bomber was used for the first time in combat. Before then the stealth program had been one of America's most top secret, deep black projects.

The B-2 was originally built to penetrate Soviet airspace and unleash nuclear weapons. In the Kosovo campaign the aircraft carried sixteen precision-guided weapons, each weighing 2, 000 pounds.

Manned by only a pilot and co-pilot, the B-2 flew nonstop from its home at Whiteman AFB, Missouri to Kosovo in thirteen hours. Virtually invisible to Serbian radar, the B-2 was able to bomb with impunity without one single U.S. combat casualty. So impressed was President Clinton with the success of the B-2 that he personally thanked the pilots and crew involved in the campaign.

The B-2 is considered the "next generation" stealth technology, following close behind the success of the F-117 stealth fighter. The B-2 can hit a target anywhere on earth with just one mid-air refueling. In Operation Iraqi Freedom, it flew all the way from its home base at Whiteman to Iraq and back. The aircraft is made of carbon graphite composite that is lighter than aluminum, but stronger than steel. On board are 136 computers performing operations more complex than those on the space shuttle. The B-2 is virtually invisible to enemy radar; its RCS is believed to be about the size of a honeybee. The aircraft can be anywhere in the world on a few hours notice and then drop a bomb within a twenty foot circle in any kind of weather. After completing its mission, the aircraft flies away undetected.

The two billion dollar bat-winged plane was unveiled to the American public in the late 1980s, around the same time the air force revealed its F-117 Nighthawk stealth fighter. For the first time since Ken Arnold saw his flying disks, the American public and the rest of the planet were officially introduced to the black world of stealth technology.

The development of stealth technology was more secret than the Manhattan Project, the top-secret program that produced the world's first atomic bomb. For decades the USAF had steadfastly denied the existence of any stealth research program let alone any stealthy aircraft. But the search for an invisible aircraft had been ongoing since World War II when the use of the new technology radar had been so crucial in the Battle of Britain. Even though the German *Luftwaffe* was much larger than the Royal Air Force, the fact that the British had developed radar gave them a huge technological advantage that proved decisive. Adolph Hitler had to cancel his plans to invade Great Britain.

Ever since World War II radar has made it almost impossible for warplanes to attack enemy targets without being detected. The Soviet Union quickly recognized this fact and began to make significant advances in radar and missile technology. As the Cold War intensified, the U.S. began to worry that it was falling behind the Soviets. The shooting down of Francis Gary Powers and his U-2 spyplane on May I, 1960 was the first time in history that an aircraft had been brought down by a surface-to-air missile. The U-2 incident showed that the United States could no longer penetrate Soviet air space and bomb key military targets deep within the Soviet Union without its planes being detected. The combination of radar and surface-to-air missiles (SAMs) proved to be a lethal combination. The U.S. needed an aircraft with the smallest possible RCS. In other words, it needed an invisible plane.

Among the most credible of UFO witnesses are the United States astronauts. Almost from the beginning of the U.S. space program astronauts were reporting unusual craft, which they apparently believed. were otherworldly space vehicles. On May 15, 1963 Major Gordon Cooper, one of the original Mercury Astronauts, rocketed into space in his Mercury capsule for a 22 orbit mission around the earth. During his final orbit he told the Muchea tracking station in Australia that a glowing greenish object ahead of him was quickly approaching the capsule. The tracking station in Muchea was also picking up the object on its radar. Major Cooper's mission had been intensely covered by the National Broadcast Company and they immediately reported his UFO sighting, but when he landed reporters were told they were not permitted to question him about it.

This was not the first time Major Cooper had seen a UFO. In 1951 he had spotted one while flying an F-86 Sabrejet over West Germany. He reported a metallic saucer-shaped disk that could fly at high altitude and out-maneuver any fighter plane. Mercury Astronaut Donald Slayton also saw a UFO in 1951:

"I was testing a P-51 fighter in Minneapolis when I spotted this object. I was at about 10,000 feet on a nice, bright, sunny afternoon. I thought the object was a kite ... as soon as I got behind the dam thing. ... it looked like a saucer, a disk. About the same time, I realized it was suddenly going away from me, and there I was, running

at about 300 miles an hour. I tracked it for a little way, and then all of a sudden the damn thing just took off. It pulled about a 45 degree climbing turn and accelerated and just flat disappeared."

While testifying before a United Nations committee, Major Cooper said: "I believe that these extra-terrestrial vehicles and their crews are visiting this planet from other planets ... Most astronauts were reluctant to discuss UFOs. I did have occasion in 1951 to have two days of observations of many flights of them, of different sizes, flying in fighter formation, generally from east to west over Europe." Cooper also testified that one of the astronauts had actually seen a UFO on the ground. In a later interview, Major Cooper revealed that U.S. radar instruments were continuously picking up strange objects that were of "form and composition unknown to us." He went on to say there were thousands of witnesses, reports, and documents of these objects, but "nobody wants to make them public."

In June 1965 astronauts Ed White (the first American astronaut to walk in space) and James McDivitt were orbiting in their Gemini spacecraft. Passing over Hawaii, they saw a strange looking metallic object with long arms sticking out of it. McDivitt reportedly took pictures of it, but they were never released to the public.

In December 1965 two other Gemini astronauts, James Lovell and Frank Borman, also saw a UFO during their 14 day mission in space. Borman reported to Mission Control that he was seeing an unidentified spacecraft near their capsule. He was told he was seeing the final stage of their own Titan booster rocket. Borman acknowledged he could see the booster rocket, but he also saw something else. Lovell reported a "Bogey at 10 o'clock" and "We have several...actual sightings." Russian cosmonauts have also. reported seeing UFOs.

NASA pilot Joseph Walker said one of his jobs was to detect UFOs during X-15 flights. In April 1962 he filmed five or six UFOs during a record-breaking 50-mile high flight. This was the second time he had filmed UFOs in flight. Walker declined to speculate about the UFOs. "All I know is what appeared on the film which was developed after the flight." None of Walker's films were ever released to the public.

It was this ongoing search for an invisible plane, one that could fly ever higher and ever faster and be made of radar absorbing materials that was responsible for the bulk of credible UFO sightings that were so prevalent during the years after World War II. The F-117 and the B-2 did not materialize overnight. There was a long history of research and development for a stealthy aircraft beginning with Reimar Horten and Jack Northrop and their flying wings. Models and prototypes were built, developed, and tested over the years in projects that were kept secret not only from the American public, but also from government and military officials. People like J. Allen Hynek who were ostensibly investigating UFOs were not given any information about the black projects. Kelly Johnson and his team built the first stealth aircraft, the SR-71 Blackbird, for the CIA in a project so secret that the Department of Defense and the USAF reportedly knew nothing about it. All of the stealth technology that the Skunk Works had developed was classified and the property of the CIA.

It wasn't until the USAF began a competition to design and build a stealth fighter aircraft that the CIA allowed the Skunk Works to reveal its history with stealth technology. In 1975 Denys Overholser, the former Director of Stealth Technology at the Skunk Works, and his team of electrical engineers began work on a top secret project called Have Blue. Overholser designed an aircraft with an angled, faceted shape that looked like a rough-cut diamond (the "hopeless diamond"). The faceted panels deflected most radar energy in multiple directions and its radar absorbing material (RAM) and composites absorbed radar energy. The result was an aircraft that not only had the lowest possible RCS, but also was practically invisible. The Skunk Works had developed a model aircraft that had the RCS of a large bird. The aircraft's triangular, arrow shape made it naturally unstable and erratic so stability was maintained by computerized fly-by-wire controls. The computers constantly made adjustments unbeknown to the pilot.

On November 16, 1978 the air force ordered five developmental models, the YF-117 As. The plane was tested at Area 51. In October 1979 Colonel Bob Jackson put together a team of elite pilots who would fly the F- 117. The pilots were sworn to secrecy and were not allowed to tell anyone, including their wives and loved ones, what they were doing. The pilots flew at night, wore pressurized suits and were known as bandits. The F-s were later organized into the 49th Fighter Wing, Holloman AFB, New Mexico and became operational by October 1983. This was years before the American public knew of the aircraft's existence. Each F-117 purportedly costs $45 million.

In 1991 the F-117 was deployed in the Gulf War. Denys Overholser who saw the beginning of Desert Storm on television said he knew instinctively that the F-117s had led the attack. In just a few weeks the F-117 Nighthawks effectively negated Iraq's multi-billion dollar integrated air defense system. Despite the presence of thousands of anti-aircraft artillery batteries and sophisticated Soviet-made SAMs, no F-117s were shot down. Striking even the most heavily defended Iraqi targets, the F-117 ranged over the night skies of Baghdad with impunity. The aircraft had an 80 to 85 percent bomb hit rate, some against targets as small as a one-yard wide vent shaft. During the Vietnam War conventional aircraft had a-hit rate of about 30 percent.

The F-117 uses laser-guided smart bombs that are carried internally. The bomb doors open for only a moment when the weapon is released. Weapons are generally released at a range of one or two miles. During the Gulf War, forty F-117s were deployed, flying more than 1,270 missions. The aircraft dropped 30 percent of all precision-guided weapons.

The successful use of the F-117 in the Gulf War sent shock waves around the world, especially in the Soviet Union. Every radar defense system in the world (including the Soviet Union) was now obsolete. The years of secret research and experimentation had paid off. The invisible plane proved to be one of the greatest technological advances of the 20th century. There did not seem to be any defense against stealth.

The unveiling of the F 117 and the B-2 was ostensibly to show the American public what it was getting for its tax dollars. All of the black projects from the flying wings to the U-2 to the Blackbird to the F-117 to the B-2 and every prototype in between had all been secretly funded, hidden away from public scrutiny **as well as congressional approval and oversight**. As knowledge of these planes began to leak out, there was criticism from many quarters about the secret funding. Bowing to the pressure, the air force finally showed off two of its stealth planes. However, there is other stealth aircraft that are still largely unknown to the general public.

F 117 Nighthawk

The Northrop YF-23A or Black Widow II was the Nothrop/McDonnel Douglas team's entry into the Advanced Tactical Fighter competition. This aircraft is a supersonic stealth fighter designed to "supercruise," meaning it can cruise supersonic without the use of engine augmentation or "afterburning." Piloted by Jim Sandberg, a Northrop test pilot, the first flight of a YF-23A took place in October 1990. The aircraft can fly at more than Mach 2, with a maximum ceiling of 65,000 feet.

An even lesser known aircraft is the AX-17 which is supposedly derived from the YF-23 A. The AX-I7 is a stealth attack plane that is also a "swing-wing" aircraft. This means the wings can sweep forward for stability when landing, but also swing back for supersonic flight. When its wings are swept back, the AX-17 looks like a perfect triangle.

The X-36 is a prototype stealthy fighter aircraft that was

developed jointly by Boeing's Phantom Works and NASA. The remotely piloted X-36 is tailless to reduce weight and drag, and has no horizontal or vertical tails. This tailless design reduces the RCS even further than other stealthy aircraft, yet the plane is believed to be more maneuverable and agile. Work on tailless flight goes back to 1989.

There is also believed to be a stealth program in Great Britain. While the United States has admitted to at least some of its stealth aircraft, the British Ministry of Defense (MoD) has steadfastly denied the existence of HALO (High Altitude, Low Observable). The HALO program is thought to be responsible for many of the triangular UFOs that witnesses in Britain have been reporting for many years. The aircraft are thought to be built at a top secret DERA (Defense Evaluation and Research Agency) facility at BAe's (British Aerospace) Warton plant in Lancashire. A HALO aircraft is often described as a small silver triangle about 30 feet long with no wings and no visible engines. It reportedly makes no sound, but will stop, hover, and then fly off at high speed.

One place where British stealthy aircraft are tested is at RAF Boscombe Down, a DERA test center near Salisbury. On a night in 1996 about 20 soldiers were taking part in a military exercise at Boscombe Down. As they were making camp for the night, a triangular aircraft appeared and hovered very low over them for several minutes and then flew off. When the group of soldiers returned to their base the captain reported the incident to his commanding officer. The CO ordered the captain not to talk about the incident and warned him that there could be serious consequences if he breached the Official Secrets Act. The surprised captain did not expect this reaction and was reportedly dismayed when he gave his men the CO's orders.

Another incident at Boscombe Down occurred on September 26, 1994. Apparently a HALO or other stealthy aircraft had to make an emergency landing. The wreckage was quickly put inside a hangar and covered with a tarpaulin. The British Special Air Services (SAS) was reportedly called out to secure the crash site. The following day, the debris was flown out on a C-5 cargo plane. The famous Rendlesham Forest encounter is probably another incident of a stealth

103

aircraft (maybe more than one) making an emergency landing.

There is evidence the U.S. and UK cooperated closely on the development of stealth technology. In his book, *UFO Revelation,* Tim Matthews describes the close, "special relationship" between the UK and the U.S. Great Britain supported the U-2 spyplane and allowed the Americans to use RAF Lakenheath in East Anglia as a base for operations. RAF pilots were also taken to Groom Lake where they received the same training as American pilots. In the 1960s a formal but secret agreement was reached between the two governments in the use and development of special materials for stealth aircraft. The British had captured an assortment of RAMs (Radar Absorbent Materials) from the Germans at the end of World War II and the Americans wanted to use these. RAF Lakenheath was also later used as a base for the SR-71 Blackbird.

Although stealth is one of the greatest military and technological advances of the twentieth century, the search for an anti-stealth technology is already on. The Australians and Russians have been working on a new technique that searches for stealth aircraft. Anti-stealth is based on the premise that no aircraft can fly through the sky without leaving behind some air turbulence. This technique looks for the air turbulence left by an invisible aircraft, much like looking for a wake that a ship or boat leaves behind on the water. However, in the "Manchester Near Miss" (Chapter Four) the two BA pilots reported an object that made no noise and left no wake, so perhaps there is already an advanced technology which goes beyond stealth.

HALO may be among these new aircraft that are not only stealthy, but are capable of flying at supersonic or even hypersonic speeds. The "old" B-2 and F-117 sacrificed speed for stealth. The new aircraft may not even need to be stealthy because they can fly so high and so fast they don't need. to defy the laws of physics and are truly Unidentified Flying Objects, planes so top secret they officially don't exist.

8. AURORA AND BEYOND

Extraordinary claims require extraordinary evidence. --
Dr. Carl Sagan
When you have eliminated all which is impossible, then
whatever remains, however improbable, must be the truth. --
Sherlock Holmes

On May 5, 1984, an alert was triggered at the North America Air
Defense Command. A USDSP (US Defense Support Program)
satellite orbiting above the Indian Ocean picked up a hot, fast object
hurtling toward earth at 22,000 miles per hour. The USDSPs are
extremely sensitive satellites used for surveillance and air defense.
Their main job is to pick up any incoming ballistic missiles. But this
object, code-named "Fast Walker," was . determined not to be a
missile. The object appeared to be on a collision course with Earth,
but suddenly it changed trajectory and curved back out into space,
stunning those who had been tracking it. A natural object like a
meteor could not have changed its direction and it also would have
been sucked in by the Earth's gravity. Fast Walker was tracked for
another nine minutes before it disappeared. Only some type of
powerful craft intelligently controlled could have escaped the Earth's
gravity and headed back out into space. It must have been an alien
spacecraft because no earthly aircraft could fly that fast and that high.
Or could it?
        Stealth technology programs were kept carefully hidden
from the American public for decades. A vast UFO phenomenon
was allowed to develop and flourish rather than reveal the secret
technology. When the F-117 and B-2 were finally revealed, their
prototypes had already been flying for decades. The revelations
were timely for both aircraft were soon deployed in the Gulf War
and then later in Kosovo and Operation Iraqi Freedom. But
according to one theory, the main reason stealth technology was
revealed was because it was already obsolete. Other, more exotic
aircraft were already taking their place and the air force was willing
to reveal the F-117 and the B-2 mainly to divert attention from its
newest deep black aircraft and protect them from public scrutiny.
        During the 1950s and 1960s research began on exotic

external-combustion propulsion systems. The idea was to build an aircraft that would be able to achieve hypersonic flight in the high Mach range by pumping fuel from its midsection into a cone of air bounded by its shock wave. Another exotic propulsion technique is called "waveriding," a vehicle's shock wave remains attached to the leading edge of the aircraft's body during hypersonic flight, giving the appearance that the aircraft is riding on its own shockwave. An unmanned hypersonic vehicle using this method of exotic propulsion would be able to operate at speeds of Mach 10 or even higher.

Exotic propulsion systems are not confined to unmanned vehicles. There is ample evidence that the U.S. has long been experimenting with manned aircraft that would be capable of flying at hypersonic speeds. These are highly classified advanced military aircraft that use pulsed detonation engines (POEs), also known as pulsed detonation wave engines. Other exotic propulsion methods include external combustion engines and the so-called waveriders.

According to the Federation of American Scientists web site, these engines "use a shock wave created in a detonation --an explosion that propagates supersonically--to compress a fuel oxidizer mixture prior to combustion, similar to supersonic inlets that make use of external and internal shock wave for pressurization." POEs purportedly have been used in top secret aircraft. Another method of exotic propulsion is the dual propulsion system.

Again, according to the Federation of American Scientists web site: *Jet engines buried in the fuselage would propel the vehicle to supersonic speeds, when a novel external burning mechanism would take control as the fundamental propulsion method. In the high Mach regime, misted fuel is ejected from the fuselage midsection - the 'break point' of the elongated diamond-- across the aft surface tiles, into the area between the fuselage and a shock wave attached to this break. In essence, the sloping, converging aft fuselage sections form the inside of a 'nozzle' and the shock boundary constitutes the outer surface, creating an expanding exhaust effect, much like that on a conventional rocket. The fuel ignited by surface heating-- or other means-- creating combustion that accelerates the aircraft up to the Mach -8 regime.*

At this point in time, there is probably no way of ascertaining just how many UFO sightings can be directly linked to glimpses of these exotic propulsion aircraft. Experimentation with these planes goes back to the end of World War II and certainly continued through the uneasy decades of the Cold War.

Aircraft that use PDEs leave a distinctive contrail, the so-called "donuts-on-a-rope" that look like a series of donuts strung on a rope or like beads on a string. These aircraft also make unusual engine noises, which some witnesses have described as a sonic boom you can almost feel; a deep, thrumming, pulsing, booming noise that almost sounds like the sky is being unzipped. Intercepted radio transmissions suggest a low frequency (about 1 Hz.) pulsing sound. Lighting patterns from these aircraft show a peculiar triangle or diamond shaped pattern.

Visual reports of these exotic aircraft go back at least to the 1980s: *Described as 'cotton balls strung on heavy yarn' by D.C. Card, a mechanical engineer who saw it forming at high altitude over Denver in 1989, this distinctive contrail matches exhaust patterns that could be expectedf rom a pulse detonation engine. At its high altitude, the unknown dark-Grey aircraft was barely visible in the late afternoon sunlight. By 1992 it was reported that "Donuts-on-a-rope contrails produced by an unknown high-speed, high-altitude aircraft have been reported throughout the U.S. and Europe, suggesting that the classified "pulser" is no longer confined to a test range ... In late January, a similar contrail described as a "coiled spring" was seen over Scotland behind a very fast aircraft flying east to west. The distinctive contrails have been spotted during daylight hours over Portland, Ore., Washington Dulles International Airport, Va., and Edwards AFB, Calif..*

In California, periodic "skyquakes" have been heard that are believed to be associated with POE aircraft: "Observers ... tell of a swift, high-altitude light that accompanies the pulsing noise ... the light moved from horizon to horizon -- well over 100 miles -- in under a minute." Another observer described the noise as a " ... very, very low rumble, like air rushing though a big tube.." Someone else said it was like" the sky itself is tearing."

On March 23, 1992 near Amarillo, Texas, Steven Douglas took a series of photos of the donuts-on-a-rope contrail. He described the engine noise as a "strange, loud pulsating roar...unique ... a deep pulsating rumble that vibrated the house and made the windows shake ... similar to rocket engine noise, but deeper. with evenly timed pulses." There is still some debate as to whether or not exotic propulsion aircraft actually exist. Those who believe they do exist are convinced they are a follow-on to the SR-7 I Blackbird.

Shortly after 9:00 PM on December 29, 1980 Betty Cash, 52, was driving through a pine forest on a deserted rural road toward her home in Dayton, Texas. In the car with her was her friend Vickie Landrum, 57, and Landrum's grandson, six-year-old Colby.

The two women say a brilliant object descended directly ahead of them, spitting flames from its underside. Cash jammed on her brakes and brought the car to a halt about 150 feet from the object. Immediately they felt intense heat inside the car and heard a loud roaring sound.

They got out of the car and stared at a blinding light and a metallic structure, big as the 200-foot-tall water tower in Dayton, and shaped like a diamond (emphasis mine) with a blunt top. The object seemed to be struggling to ascend above the treetops, emitting blasts of fire and a continuous roar that reminded the women of a shrill welding torch, only much louder.

After ten minutes, the object rose above the trees, tilted itself onto one side, and began moving slowly south. Then, by their account, up to 23 helicopters eventually appeared, apparently following the object.

On the drive home, Landrum, Cash and Colby said they suffered headaches and later nausea, and over the next few days experienced bouts of vomiting, diarrhea and skin burns. Cash also lost large clumps of hair, and on January 2, 1981she was admitted to Houston's Parkway General Hospital as a burn patient. She spent four of the next five weeks under supervised care.

According to Randy Fitzgerald, writing in the *Reader's Digest,* Cash and Landrum contacted various authorities, but no one could give them any explanations. John F. Schuessler, a NASA contractor, and Alan Holt, a NASA physicist, interviewed the two women and were taken to the site of the encounter. The highway's yellow line had wiggled from the melting of the heat blast and a 20-

foot circle of the road surface seemed to have melted and then resolidified. About 20 feet above ground level, the trees facing the road were blackened. Schuessler interviewed five witnesses who had seen a similar UFO that night. Another eight witnesses, including a Dayton police officer, saw the helicopters, but not the UFO itself. Schuessler spent the next five years trying to find and identify the helicopters, but without success. The incident was also investigated by an Inspector General from the Department of the Army, but he offered no conclusion.

Another UFO case that seemed to involve a close encounter with a radioactive craft occurred on April 1967 in Jefferson City, Missouri. The witness was a school principal, and the sighting was listed as Unidentified in the Blue Book files.

The incident was reported by J. Allen Hynek in his book *The UFO Experience:* "*I was coming home from a PTA meeting and heading down a small country road .. All of a sudden I noticed a glow coming from over the cliff. .. Then this unbelievable object -- shaped something like a World War I helmet - came over the top of the cliff. . .l slowed down. .. and this huge object, over 300feet, I'd estimate, came over the cliff and stood still almost directly over me for a split second like any object changing direction and then took off toward the airport .. .It was terrific bright light. The top of the car seemed to have no effect in holding out the light. It was a terrific bright light, unbelievable ... When I looked at my hands, it looked like I was looking at X-ray photos.* "

Other people also witnessed the Jefferson City UFO. After his close encounter, the school principal drove to the airport to report what he had seen. "But I didn't have to tell the people inside. They'd seen it." When the air force came to investigate, the attitude of the investigating lieutenant "was not Did you see it? but How much of it did you see?"

The Aurora is one of the most secret of deep black aircraft In a 1985 Pentagon document, the code-name" Aurora Project" was accidentally included, but there has been no official mention of it since then. In his book *The Demon-Haunted World* the late Carl Sagan wrote of the Aurora:

Aurora is *a high-altitude, extremely secret American*

reconnaissance aircraft--a successor to the U-2 and the SR-71
Blackbird. It either exists or it doesn't. By 1993, there were reports by
observers near California's Edwards Air Force Base and Groom
Lake, Nevada, and particularly a region of Groom Lake called Area
51 where experimental aircraft for the Department of Defense are
tested, that seemed by and large mutually consistent. Confirming
reports were filed from all over the world. Unlike its predecessors, the
aircraft is said to be hypersonic, to travel much faster, perhaps 6 to 8
times faster, than the speed of sound. It leaves an odd contrail
described as donuts-on-a-rope. Perhaps it is also a means of
launching small secret satellites into orbit, developed, it is
speculated, after the Challenger disaster indicated the shuttle's
episodic unreliability for defense payloads. But the CIA "swears up
and down there's no such program, " says U.S. Senator and former
astronaut John Glenn.

The principal designer of some of the most secret U.S.
aircraft says the same thing. A Secretary of the Air Force has
vehemently denied the existence of such an airplane, or any program
to build one. Would he lie? "We have looked into all such sightings,
as we have for UFO reports, " says an Air Force spokesman, in
perhaps carefully chosen words, "and we cannot explain them."
Meanwhile, In April 1995 the Air Force seized 4,000 more acres near
Area 51. The area to which public access is denied is growing.
Consider then the two possibilities: that Aurora exists and that it does
not ... Either way, Aurora seems relevant to UFOs.

In 1979 it was reported that a high-performance piloted stealth
reconnaissance aircraft was being developed to replace the SR-71
Blackbird. This aircraft would be able to fly in the hypersonic (Mach
4 - 6) range with a 200,000-foot ceiling, the very edge of the
stratosphere where the mesosphere begins. It is believed the Skunk
Works was involved with this project which had been underway at
least since 1987. A first flight was thought to have occurred in 1989.
The total development for the Aurora program is believed to range
from $4.4 billion to $8 billion. A fleet of 24 to 30 aircraft would cost
about $10 billion to $24 billion.

The Federation of American Scientists reports that the Aurora
is being flown from a base in Area 51 to a Pacific atoll and then on to

a remote RAP airbase at Machrihanish, Scotland for refueling before it returns to the U.S. at night. The Aurora can also be refueled in mid-air by specially modified tanker aircraft that use liquid methane fuel. After refueling in Scotland, the Aurora is believed to fly across the North Pole at speeds as high as Mach 8 (5,300 mph) on its way home. As it lands, a squadron of F-111 fighter-bombers flies in close formation to confuse civilian radar. With a two-man crew, the aircraft can fly so fast and so high it is invulnerable to any current missile system. The Aurora is "beyond stealth," meaning it can fly over a target so fast and so high it doesn't need stealth technology.

Another clue to the Aurora's existence are the so-called "skyquakes" that have occurred on a regular basis in southern California. A skyquake is a boom so powerful that it can be detected by earthquake monitoring equipment. Aircraft associated with these skyquakes have flight paths, which show a direction toward Groom Lake in Nevada. The RAF has also reported skyquakes over northern Scotland.

The Aurora is probably powered by a POE which is the most powerful engine currently known. It can produce more thrust than any other known engine. The contrails seen in the wake of the skyquakes are the classic donuts-on-a-rope, and the only engine, which leaves behind this type of contrail, is the POE. Interestingly, POE engines are able to perform the so-called "hyperjump" maneuver. This allows the aircraft to accelerate at a speed greater than the human eye can follow. An observer who saw the maneuver would think the aircraft had disappeared. Perhaps this explains the accounts of all those UFOs that suddenly and inexplicably disappeared.

The Aurora is painted black and is triangular or diamond in shape. It is a large plane with some estimates as long as 100 to 200 feet. The aircraft is most likely made of titanium and carbon composites in order to withstand the tremendous heat build up from flying at hypersonic speeds. An aircraft flying between Mach 5.5 and Mach 6 would have a skin temperature as high as 1100 to 1300 degrees F. Flight crews on the Concorde said you could put your hand on the plane's inside wall and feel the heat, and Concorde only flew at Mach 1 to 2.5.

Witnesses who may have spotted the Aurora at night typically report a single, bright light, sometimes pulsating, flying at a speed that far exceeds other aircraft in the area and flying at altitudes estimated at 50,000 feet or more. There is often no engine noise. Sometimes the aircraft seems to be hovering silently. Other observers report a distinctive triangular or diamond shaped lighting with two white lights near what would be conventional wingtips and one amber light near the tail and sometimes a green light near the nose. Although the two white lights are reported to be much brighter than lights on a conventional aircraft, they do not illuminate the Aurora's platform. Witnesses have claimed the aircraft's wing lights are twice as far apart as those are on the F-117 and the nose-to-tail lights are about 50 percent longer than those on the stealth fighter. Sometimes a low peculiar rumbling noise is heard. Witnesses who see the diamond lighting pattern and hear the low thrumming noise are probably seeing the aircraft on its landing approach.

Another futuristic aircraft is the Black Manta TR3-A. Although not as highly advanced (but just as highly classified) as the Aurora, the Black Manta is thought to be part of "Team Stealth." The aircraft gets its name from its shape which resembles a manta ray. While the hypersonic Aurora would fly quickly over a target, the Black Manta would be a low-speed stealth aircraft that would hover over a target for a long period of time in order to relay information back to an attack or bomber aircraft like the F-117 or B-2 or even a British HALO. It is believed the Black Manta performed this role in the Gulf War. The aircraft is also able to illuminate a target with a laser that is used to guide bombs. It was probably a "Team Stealth" exercise that was responsible for the Belgian UFO flap (Chapter Four). In 1991. Lockheed reportedly tried to get funding to build and modify an extra 24 F-117s. It's possible this might have been a way of getting funding for the Black Manta program without revealing the actual aircraft. Holloman AFB is thought to be its home base.

On the night of March 13, 1997 from around 7:30 to 10:30 P.M. hundreds of people including state troopers, airline pilots, military officials, and ordinary citizens saw a light display in the skies over central Arizona that has been dubbed the "Phoenix Lights." Circular orange lights were seen silently moving along a path from Las Vegas to Phoenix. Some witnesses described the UFO as absolutely huge, anywhere from one to three miles wide, flying silently and hovering.

According to the CNN web site, Tim Ley and his family were among those hundreds of witnesses who saw the lights. The Ley family first saw the lights while they were looking north from their home in Phoenix. Said Ley, "When it finally got here and we realized this thing was coming right over us, we really started getting antsy." Then he and his son Hal watched it fly directly overhead in complete silence. Ley said that "when the right side of what appeared to be a giant V -shaped craft passed directly over him, the left side was a couple of blocks away." His wife Bobbi said the size of the craft was overwhelming, but "It didn't seem threatening ... When it was right overhead and we couldn't hear a sound, it was like you're just awestruck." .

Jim Dilettoso of Village Labs, Inc. in Tempe, Arizona described the various manifestations of the lights. At one time they were observed in a hexagram pattern over the Superstition Mountains. The formation of amber orbs was next seen in two separate arc patterns with "trailing lights" over the Gila River area. The last appearance of the hovering orbs was in a V-formation over Rainbow Valley. Another described a high speed V-formation of five to seven orbs, seen as bright blue-white to yellow-white. This formation was seen leaving the Henderson-Las Vegas area and then coming through the Prescott Valley at a low altitude. When the craft slowed down, the orbs at the point appeared as amber colored. Observers in the Oracle area reported the color as red to red-orange before the formation split up.

Dilettoso described a "Chevron," a huge, single object that was first described as a V-formation of five white lights moving slowly and silently toward Prescott Valley. This was probably the object seen by the Ley family. "The huge craft made a pivot turn over the witnesses' home while the lights shifted into an arch formation while turning to red, and upon doing so, it suddenly shot off at blinding speed toward Phoenix." The next witnesses saw the UFO coming directly toward them at an estimated speed of Mach 2 or Mach 3. It then changed back into the Chevron and cruised slowly and silently directly over their heads. Other witnesses along its flight path later saw the craft. Its size was enormous, estimated at 900 feet or more on each leading edge and more that a mile in length to its tail. At Picacho Peak, it hovered for about ten to fifteen minutes until the witnesses lost sight of it. Some fifteen minutes later a "giant triangle" followed its path down I-10.

It's hard to imagine an earthly craft as large as the one described by witnesses on that night in Arizona. The lights were recorded on film and video by several amateur photographers along a 300-mile corridor from the Nevada-Arizona state line through Prescott and Phoenix to the edge of Tucson. The footage was eventually shown on worldwide television. "Arizona Invaded By UFOs," trumpeted one newspaper in Britain. Throughout the state, police stations and military bases were fielding frantic calls from terrified citizens who demanded an explanation. Maybe this really was the long anticipated alien invasion.

Lame explanations ranging from low-flying aircraft to the dropping of flares offered by military and government officials were eerily reminiscent of the cover-ups of past decades. It sounded like temperature inversions and weather balloons all over again. One official who tried to find an explanation was Phoenix City councilwoman Emma Barwood. She spent months trying to get at the truth, but was constantly rebuffed and sometimes ridiculed.

She said, "The issue is not whether or not they were flying saucers. The issue is why the government refuses to tell the American public what, exactly, those lights were. " Unable to get the truth from the government, a frustrated Barwood decided to run for secretary of state of Arizona. Her platform included a demand for a serious investigation of the lights, and she promised to solve the mystery if elected. "If the lights were anything else, from an Iraqi invasion to an extraterrestrial visitor, the government needs to let us know, if for no other reason than to prevent our imaginations from running rampant." She also said, "From what I have learned about this in the last year or so, things like this have been going on all over the world and have been for some time. "

When asked why the government won't level with the public she responded, " .. .1 believe it's for one of two reasons. They know what the lights were, but for some reason they don't want anyone else to know. Or, either they don't know what they were, or they don't have any control over them." Barwood even sent Senator John McCain of Arizona two letters requesting information about the Phoenix Lights. The senator reportedly sent several letters to the USAF searching for answers. But, the air force's position was that it stopped investigating UFOs in 1969 when Project Blue Book was officially closed.

So what could account for a silent craft perhaps over a mile long? Is there an explanation other than an extraterrestrial one?

The most plausible answer is that the Phoenix Lights belonged to a hybrid, lighter-than-air (LT A) craft that is part aircraft and part airship. Experiments and test flights of such an aircraft purportedly go back to the late 1960s. During that period, the Aereon Corporation in Princeton, New Jersey was a leader in the design and development of hybrid aircraft. One of their first designs, the Aereon 3, harkened back to the cigar-shaped blimps of the 1930s. This design was abandoned in 1967, and was replaced by the "deltoid pumpkin seed" that used helium cells that were integrated into its hull. The 27 foot-long Aereon 26 prototype was tested in 1971.

The next craft developed by the company was the Aereon 340. This craft looked like a large triangle with a length of 340 feet and a.wing span of 256 feet. It was powered by four turbo-prop engines as well as helium gas. Its purpose was to be for transporting cargo. During the 1970s the military became interested in hybrid aircraft as means of oceanic surveillance. The result was the "Dynairship," a huge triangle 600 feet long and with the capability of remaining in the air for days and hovering over the water. But it could also reach speeds of over 200 mph if necessary.

One of the Phoenix Lights manifestations described by Dilettoso was an enormous craft with a definite structure complete with panels in a grid pattern along with lights at each of the three corners and a larger center light. The only explanation seems to be hybrid aircraft. " ... More and more people are coming to the conclusion that the larger triangular aircraft seen in our skies, especially those that appear to have pipework patterning on their underside are helium-filled LTA vehicles."

Hybrid aircraft seem to be the most logical explanation for the Phoenix Lights as well as for some of the Hudson Valley UFOs. Such aircraft would be large deltoids, would tend to operate at low levels over different terrain, and have the ability to hover or fly at low speed. Their near silent operation could be accomplished by ducted turbo-fans or a similar propulsion system.

Another possible hybrid aircraft reported in Tim Matthews' book comes from California. In the late 1980s, witnesses saw a huge, black flying wing that stopped, hovered vertically and pointed its trailing edge toward the ground. Such a bizarre maneuver could be effected by a LTA craft being pushed by slow-turning propellers. Hybrid aircraft could be used for surveillance, as a "mothership" for transporting or launching smaller aircraft, as a means of secretly depositing troops behind enemy lines, as atmospheric satellites, sensor platforms, or for use as large radar platforms.

The so-called Silent Vulcan seen in the UK may also be a hybrid aircraft: It fits the descriptions.

I was traveling home and I was between Littlethorpe and Cosby at 6.20 PM when I noticed 3 white lights travelling towards me. I thought at first it was an aeroplane going to land ... As it got nearer, the lights got very bright and formed a triangle, this in itself seemed an unusual formation. I slowed down to view it more safely and realized there were no other lights as on normal aircraft .. .! stopped my car and opened the windows as it got closer. I could see the shape as a black silhouette, it was like a Vulcan bomber, triangular shaped with a light on each point and large. The other strange thing was that I heard no noise at all, no sound of engines etc. -- after it passed, the lights were no longer visible and it disappeared from my view into the dark. I have never seen the like before and that includes three years in the RAF when I saw aircraft day and night. (Mr. K. Porter.)

And from Mr. Mayne: *Being ex-RAF I would have said the object was a Vulcan bomber flying South to North on the flightpath used by the East Midlands airport. However, the object was stationary for 3 minutes. Imagine a Vulcan with bright lights on each wing tip and two very bright landing lights in the centre. It then moved towards us and overhead, (then) gathered speed. When overhead I could see what looked like a red navigation light in the centre and another white light where the tailplane would be. THERE*

WAS NO NOISE. I rushed to the front of our bungalow ... and it had completely disappeared.

These two sightings were reported in Leicestershire, UK in October 1978. Between 6:00 and 11 :46 P.M. on that night, Leicestershire's UFO groups were inundated with telephone calls from people who saw a large, triangular object with three or four bright lights on its underside. The object's characteristics -- large size, silent operation, slow movement and unusual lighting patterns -- suggest this could have been an early version of a hybrid LTA craft. By 1978 the Aereon Corporation had carried out successful test flights of their deltoid LTA craft and had produced several designs for a large triangular airship.

Another futuristic aircraft has been seen cruising over the California desert. The FAS web site calls this craft "Mothership." It is reportedly large and agile with a wingspan of 150 feet, similar to the B-2. But unlike the B-2, this aircraft seems to be highly maneuverable, able to turn 90 degrees on its wingtip. It is described as a large, light-colored aircraft with a platform similar to the XB-70 supersonic bomber. It has been seen a total of five separate times by residents of Mojave, California and by workers at Edwards AFB:

Observers said they first saw a large, primarily delta shaped aircraft at night during the summer of 1990 ... Engine noise associated with the aircraft seen on Sept. 19 was described as a low-pitched rumble. Afterburner flames from twin exhaust ports located under the wing trailing edge and immediately outboard of the aircraft center line during the Oct. 3 sighting. .. A similar aircraft was seen in April, 1991 at about 11 a.m., flying north of Edwards AFB at an estimated altitude of 5, 000 to 10,000 ft. An observer said it was large -- dwarfing an F-16 chasing it -- and was light colored, possibly white.

Mothership is apparently a large vehicle, possibly 200 feet long. It has light-colored top and bottom surfaces, with dark leading and trailing edges. It is called "mother ship" because it is believed to be used for launching a smaller, air-launched hypersonic aerospace plane.

In 2001 NASA unveiled its X-43A, an unmanned hypersonic aircraft designed to fly seven to ten times the speed of sound. The

117

aircraft is 12 feet long and shaped like a surfboard. It has an air-breathing engine, which means it carries hydrogen for fuel, but must scoop oxygen out of the atmosphere to combust it. For this supersonic combustion ramjet or scramjet engine to work, the X-43A needs a rocket boost. A B-52 will haul the plane to about 24,000 feet then release it. A booster Pegasus rocket will ignite and accelerate the X-43A to its hypersonic speed and its altitude of about 100,000 feet. The aircraft will then separate from the booster and fly under its own power. Unlike normal jet engines, where rotating blades do the compression work, the scramjet works by air flowing through it at supersonic speeds.

Deep black aircraft require special, secret bases that can not only maintain the aircraft's secrecy through tight security and restricted access, but can also provide the special maintenance facilities these aircraft require such as specially long runways. Because these aircraft fly at hypersonic speeds, they generate tremendous heat build-up and need to be put into special cooling hangars after landing. They also have special fuel requirements like liquid oxygen. Groom Lake (Area 51) and Edwards AFB are the two main bases for secret aircraft in the United States. This is probably one reason why so many UFO sightings come from the American West including Arnold's original sightings.

Other secret bases can be found in the UK, which indicates not only the "special relationship" between the two countries, but also the very real possibility that Great Britain has its own deep black aircraft programs.

RAF Macrihannish is located on the tip of the Kintyre Peninsula on the West Coast of Scotland. The U.S. SEALS Special Forces used this base, and in the 1980s was a test center for the F-117. Witnesses living in this area have a history of reporting UFO sightings ... and crashes! More recently, loud noises and unusual contrails have led to the conclusion that this base is involved with Aurora. Indeed, a Scottish base is believed to be involved in the Aurora's refueling before it flies over the Arctic Circle on its way back to its home base in the U.S. (probably Edwards AFB). Machrihannish is purportedly closed and under "Enhanced Caretaker Status."

RAF West Freugh, also on the West Coast of Scotland, is

believed to be a test center for BAe's HALO project. RAF Boscombe Downe has long been a home for experimental aircraft, and has been involved in some of the more famous UFO incidents in the UK. It is located near Salisbury, England.

Aircraft like the Aurora did not just materialize overnight. They have a long history that reaches back more than fifty years. On October 2, 1942 test pilot Bob Stanley completed the first official flight of the Bell XP-59A, America's first jet, the Airacomet. Like all the black projects that came after it, this too was a top-secret program.

Another deep black project was the X-15. This aircraft was made from titanium in order to tolerate the 1200-degree F heat generated by hypersonic flight. Flying faster than Mach 5, it could climb more than 50 miles above the Earth's surface. This dart-like aircraft was the first plane to be piloted into near space.

Perhaps Edwards Air Force Base's own web site sums it up the best: *"Down through the years, a fascinating variety of aircraft have been associated with Edwards AFB, and the Air Force Flight Test Center. Mostly futuristic, occasionally unique, but always highly advanced, these planes have consistently pushed the boundaries of flight performance and expanded the capabilities of the aircraft now flying today ... Very often, our test vehicles have dramatically smashed the limits of airspeed and altitude... Airframes and systems are continually being evaluated in order to expand and improve the performance levels of modern aircraft in every way imaginable: maneuverability, reliability, time-to-climb improvement, safety, cargo-carrying ability, navigation, takeoff-effectiveness, and numerous other flight and ground parameters. Inevitably, this ongoing effort to push the limits of performance has resulted in some fantastic and unique shapes winging through the Edwards airspace. Some, which were unbelievably advanced for their day, now seem almost quaint alongside modern aircraft. Others achieved performance levels which have never since been surpassed. Some looked almost bizarre, being designed for a single, specific purpose ...*

Many ufologists believe highly advanced aircraft have been "back engineered" from captured UFOs. Nothing made on Earth, they argue, can perform like these aircraft. Groom Lake is believed to be the base where the captured alien spacecraft are kept, studied,

119

and tested.

LoFlyte (Low Observable Flight Test Experiment) is an experimental aircraft that uses a neural network. This is a computerized flight control system which actually learns by doing. The aircraft's flight control system learns as it flies. The neural network system is designed to help pilots control an aircraft that will be flying hyper-sonically in the Mach 5 or more range. The neural network will also help pilots land a damaged aircraft even if the controls are partially disabled.

A neural network computer system has many interconnected control systems called nodes which are similar to neurons in the human brain. "Each node assigns a value to the input from each of its counterparts. As these values are changed, the network can adjust the way it responds. The LoFlyte aircraft's flight controller consists of a network of multiple-instruction, multiple-data neural chips. The networks will be able to continually alter the aircraft's control laws in order to optimize flight performance and take the pilot's responses into consideration." Eventually the neural network system could learn to control the aircraft.

The futuristic diamond-shaped aircraft uses the wave rider concept: a hypersonic vehicle that cruises on top of its own shock wave. The wave rider principle uses shock waves built up during hypersonic flight to increase lift. "The wave rider surfs on a stream of air because of the wedge-shaped design of the aircraft. The sharp leading edges of an aircraft shaped like the LoFlyte tend to wrap the shock waves around the fuselage. A cushion of air is trapped below it." Terrence Nonweiler, a British scientist, proposed a wave rider system as far back as the 1950s. In 1951 he published his first paper on wave riders, and by the end of the decade was working on the "theoretical aspects of a manned re-entry vehicle." NASA and Accurate Automation Company of Chattanooga, Tennessee developed LoFlyte.

If you want an aircraft that looks like a classic 1950s flying saucer then you have it in Lightcraft. The Lightcraft vehicle uses "beamed-energy propulsion" and is designed for both transportation here on Earth as well as space flights. (20) The Goodyear blimp inspired its lightweight design (there is no on-board propellant). Propulsion is provided by three different modes: ion-propulsion mode, Magnetohydro Dynamic (MHD) propulsion mode, and Pulsed Detonation Engine mode (PDE). The vehicle can take off vertically and then accelerate into space. The on-propulsion mode is the only on-board propulsion system and is used only for low-speed transport. Its use is mainly for landing and taking-off. Speeds in this mode are equal to a train. Solar cells, which cover one side of Light craft, run this system.

PDE and MHD modes are powered by beamed energy from anorbiting satellite with laser and/or microwave transmitters, but can propel the vehicle much faster than the ion mode. The MHD engines accelerate it to orbital speeds within Earth's atmosphere and are required for orbital flights. According to the Lightcraft Project at Rensselaer Polytechnic Institute website, "The MHD engine system includes a special 'expendable coolant' system that efficiently transfers waste heat from the rectennas to on-board de-ionized water, which is subsequently vaporized and ejected as steam."

To get Lightcraft into space, the PDE engines perform a hyper jump maneuver. After reaching supersonic speeds, the MHD engines come on line and perform the final push into space. For landing, the vehicle can use an auxiliary tripod landing gear, three or more extended pods, or use the Maglev Lander.

At present, Lightcraft is in its testing phase under the direction of Leik Myrabo. The tests are held at the Army's High Energy Laser Systems Test Facility (HELSTF) near the southern end of the White Sands Missile Range. The test vehicle is a palm-sized structure that looks like a flying saucer. It is made of aircraft aluminum and has no moving parts. Future models will probably be built from composites or high-temperature ceramics. It was originally part of the Star Wars anti-missile research. The underside of the test vehicle is a parabolic mirror that focuses the laser toward the edge. Exploding air that reaches temperatures from 10,000 - 30,000 degrees Kelvin provides thrust (18,000 - 54,000 degrees F). When launched, there is no chemical exhaust.

9. A GENUINE UFO EVENT?

*It is the policy of the FAA that UFOs do not exist, so we do not collect
any reports of such.*
-- In *Night Siege: The Hudson Valley UFO Sightings* by Dr. J.Allen
Hynek and Philip Imbrogno

In the annals of UFO lore there is one incident in particular that is
officially classified as a "genuine UFO event" (1) The incident
occurred on October 4, 1967 around 11:00 PM in the small fishing
village of Shag Harbor about 125 miles from Halifax, Nova Scotia. It
began when two men, David Kendrick and Norman Smith, saw some
unusual lights above the tree line. Many other villagers also saw the
strange lights, which consisted off our bright lights that flashed in
sequence. Then suddenly, without warning, the lights tilted to a 45-
degree angle and plunged into the water. As the object impacted, there
was a bright flash and loud roar. No mention was made of a UFO. The
local villagers simply assumed a large aircraft had crashed into their
harbor.

The Royal Canadian Mounted Police were notified. Corporal
V. Werbicki and Constable Ron O'Brien were dispatched to the scene.
Constable Ron Pond who had been on highway patrol had actually
seen the object while it was still in the air. The unusual lighting
patterns and strange flight characteristics of the object had already
concerned him, even before he heard from the other officers. As the
RCMP officers hurried to the scene, a group of villagers had gathered
by the shore and were watching a pale yellow light moving on the
water. As the people watched incredulously, the light seemed to
disappear. Local fishermen, still assuming a large airplane had
ditched, got into their boats and headed for the spot, hoping to rescue
any survivors.

When the three RCMP officers arrived they could still see a
pale yellow glow floating on the water about a half-mile from the
shore. The object was leaving a trail of dense, yellow, bubbly foam
that drifted in on the tide. Constable Pond said the object had
changed shape before it plunged into the water and that it "appeared
to be no known object." The yellow foam continued to bubble to the
surface and a large slick developed. Search efforts continued until

early morning and then resumed again during the day. A check with the Rescue Co-ordination Centre in Halifax and with the nearby NORAD facility at Baccaro, Nova Scotia confirmed that no aircraft, either civilian or military, had been reported missing. Yet everybody involved in the rescue mission was convinced that something real and solid had crashed into the water.

The next day the RCMP made contact with nearby military bases. The incident had occurred near Government Point, a secret U.S. base on Canadian soil (The Shelburne Military Base is now abandoned). A Canadian Navy diving team was dispatched to look for debris on the harbor bottom. Because an official statement said they were looking for a UFO, the incident quickly became a headline story, but after no debris or other corroborating evidence was found the story gradually died away.

The Shag Harbor event became Case #34 in the Condon Report. Dr. Levine, a member of the committee and a research engineer at the University of Arizona, was assigned to investigate the case. Maritime Command and a RCMP spokesperson told Dr. Levine that there was nothing to the incident and no reason for any further investigation.

Twenty-six years later the case was reopened as more information began to emerge and men from the original Canadian Navy Diving Team spoke up. Canadian ufologist Chris Syles began to sift through the large assortment of official documents related to the case. All of this painted a picture of a genuine UFO event.

First of all, it was the authorities that said that the object was a UFO. None of the original village witnesses had ever called it that. All references to a strange object that behaved in an unusual manner came from official sources. The large amount of documentation revealed a serious search effort was made to locate the UFO which officials apparently felt was of extraterrestrial origin.

According to the navy diving team, the UFO had been detected on the harbor bottom, but got away from them and went up the coast to Government Point. Although the Shelburne Military Base is now abandoned, it was once the first line of defense against a Soviet attack. Here American forces continuously monitored the sea for Russian submarine activity using a sophisticated underwater grid.

While the UFO was in the water off Government Point, six to seven navy ships supposedly stayed over it for a week while another UFO repaired it. While these repairs were being made a Soviet submarine, in violation of the then 12-mile international limit, showed up and attempted to make contact with the underwater UFOs. But, before the confrontation became serious the two submerged UFOs moved away down the Nova Scotia coast and into the Gulf of Maine. Once in the Gulf of Maine, they reportedly surfaced and flew away. Their flight prompted more UFO reports.

Unlike most UFO incidents this one has a beginning, a middle, and an end. The object was spotted, it crashed into the harbor, it moved down the coast where it then surfaced and flew away. In addition, numerous officials and government agencies had been involved and there were official records documenting the event. Also, this particular UFO was not just a brief glimpse of an unusual object. It was around for a period of seven days from October 4- I I. While it was being repaired, it was seen and monitored by military personnel. Apparently, it was even detected by a Soviet submarine, which tried to investigate. It was officially classified as a UFO.

Shag Harbor was not the first underwater UFO. A previous incident occurred on October 31, 1963 on the banks of the Peropava River in the Sao Paulo province of Brazil. Like Shag Harbor, witnesses initially reported seeing a bright light moving through the treetops.

The light then reportedly wavered, changed course, and began rocking violently. It moved out over the river and then abruptly plunged into the water. Mud and debris came bursting to the surface along with an explosion of large bubbles. This is reminiscent of the bubbly foam that would be seen four years later in Shag Harbor. The Brazilian fishermen who had seen the light described a disk-shaped object, about three feet thick and between 15 and 20 feet in diameter (Witnesses said the Shag Harbor object was about 60 feet in diameter). They said it shone brightly, like polished aluminum. Interestingly, the fishermen thought it was not manned but was being operated by remote control.

Just as it would later be in Shag Harbor, this incident became a media event. Every newspaper in Brazil reported the story and hordes of sightseers and UFO investigators descended on the Peropava. On November 2, a local diving instructor along with two colleagues searched the river and tried to find the disk but found no sight of it. Over the next few days, several attempts were made to find the disk, but they all ended in failure. Some thought the disk had either disappeared or moved downstream, underwater and under its own power. Despite the widespread public furor over the incident, the Brazilian government and military authorities showed no interest in the UFO and made no official investigation or comment

Over the years Brazil has had its fair share of UFO reports. One of the most famous occurred on January 16, 1958 when a geologic survey team saw a strange object flying toward Trindade Island in the South Atlantic. The object was captured on film and for decades this incident was regarded as one of the most credible UFO sightings. The still photographs show a round object encircled by a flat disk. Viewed today, it clearly looks like a stealthy aircraft, probably an early prototype.

Both Shag Harbor and the Peropava River incidents were probably encounters with unmanned vehicles that were part of an Underwater Unmanned Vehicle (UUV) system. These programs are underwater surveillance and reconnaissance systems that were and still are highly classified. Very little information is available on them, but the programs include vehicles like the Manta Program which is part of an Advanced Unmanned Underwater Vehicle (AUUV) surveillance system. UUVs and AUUVs could also be part of highly secret and highly sophisticated underwater defense systems. Boeing has been involved in underwater systems for at least 30 years. This fact, coupled with the Shag Harbor UFO moving up the coast to a secret U.S. military base, would indicate that the object was very likely part of a 1967 version of a top secret underwater defense system. Maybe that's why the Soviets sent their submarine to investigate.

Although the Shag Harbor UFO went into the ocean and the Peropava disk went into a river, there are enough similarities between the two incidents to suspect that the Peropava disk was also part of an unmanned underwater defense system. The Peropava "disk" might have been a test of an early UUV system that went haywire. Like the

Shag Harbor UFO, the Peropava disk apparently crashed into the water. Both objects then seemed to get away while still submerged. Maybe the Peropava disk disintegrated upon impact although no debris was found. No debris was found in Shag Harbor either. Perhaps the Brazilian authorities showed no interest in the incident because they knew in advance that such an underwater vehicle was being tested.

It's hard to know with certainty what role these underwater systems played in the UFO phenomenon. Some of these unmanned vehicles might be capable of both aerial and underwater reconnaissance and defense -- a Manta-type vehicle that can perhaps fly through the air and also zip around underwater. Unmanned vehicles offer many more possibilities and capabilities than manned vehicles, and their role in UFO sightings could be very. significant.

10. The LEGACY OF BLACK PROJECTS

The Truth is Out There. -- The X-Files

For five decades the United States government and military misled, disinformed, and lied to the American people about UFOs. From the very beginning, the CIA was involved in the designing, building and testing of black aircraft, which they went to great lengths to conceal. Critics charge that a UFO phenomenon was allowed to grow in order to protect the secret aircraft. The pilots who flew the first black jets resigned from the USAF and went to work for the CIA. When people claimed they had seen flying saucers, no one from any government agency came forward with even a hint of the truth. Beginning with Kenneth Arnold, credible witnesses, serious UFO investigators and ordinary people were discredited and often made to look like fools. Reputations were tarnished and sometimes even ruined. Lives were changed forever. UFO chases, either on the ground or in the air, were often dangerous and occasionally resulted in fatalities. "Official" investigators were kept in the dark about the black world and the truth withheld from them. When their investigations were completed and their final reports written, the results were usually disappointing and unsatisfactory. None of these projects and reports ever suggested

UFOs might be top secret aircraft. Why not?

In the Project 1947 report one mention is made of the "possibility that the objects could be unidentified and unconventional types of aircraft," but there is no follow up. The Colorado Project's Edward Condon briefly considered the possibility of a government cover-up and then just as quickly dismissed it. He wrote:

*Among those who write and speak on the subject, some strongly espouse the view that the federal government really knows a great deal more about UFOs than is made public. Some have gone so far as to assert that the government has actually captured extraterrestrial flying saucers and has their crews in secret captivity, if not in the Pentagon, then at some secret military base. We believe such teachings are fantastic nonsense, that it would be impossible to keep a secret of such enormity over two decades, and that no useful purpose would be served by engaging in such an alleged conspiracy of silence. One person with whom we have dealt actually maintains that the Air Force has nothing to do with UFOs, claiming that this super-secret matter is in the hands of the Central Intelligence Agency, which, he says, installed one of **its own agents** as scientific director of the Colorado study. (Emphasis mine) This story, if true, is indeed a well kept secret. These allegations of conspiracy on the part of our own government to conceal knowledge of the existence of 'flying saucers' have, so far as any evidence that has come to our attention, no factual basis whatever."*

It is worthwhile to take a close look at this excerpt from the voluminous University of Colorado Project report. First, Condon ridicules the notion that the government has captured extraterrestrial flying saucers and is storing them at a secret military base. Most reasonable people would agree with him that this is indeed "fantastic nonsense." But Condon doesn't bother to ask why some people believe this. Is it because they might have seen strange-looking craft making sharp turns and fantastic maneuvers that seem to defy the laws of physics? Such a possibility doesn't seem to occur to him.

He writes that it would be "impossible to keep a secret of such enormity." Yet in 1943 Condon helped J. Robert Oppenheimer recruit the scientists who would work on the Manhattan Project, a project so top secret not even Vice President Harry Truman knew about it. Condon thinks the possibility of a CIA operative being planted in his project is ridiculous, and he appears to be totally unaware that the CIA was the prime mover and shaker behind black projects and had a vested interest in keeping them concealed. Apparently, Condon trusted his government and confidently dismissed any notion that his panel had been compromised.

Not only did Condon abandon the notion of a government cover-up but he also made light of any "conspiracy hypothesis," pooh-poohing the idea that there might be a "supersecret government UFO laboratory hidden away somewhere" or that his project might "merely pretending to be a study of UFOs," or that his committee was really a cover story for conducting a "top secret study for the Defense Department's Martian Invasion Defense Program (MIDP), that is, a war plan for a response by our defense forces in the event of an invasion of Earth by the Martians." One can just picture the Project Director shaking his distinguished head at such fantastic nonsense. But, the central question remains unanswered to this day: Why did the Condon Committee refuse to even consider the very real possibility that UFOs might be top secret aircraft, especially when Condon himself said, "I'm almost inclined to think such studies ought to be discontinued unless someone comes up with a new idea on how to approach the problem ... " Tackling the UFO problem from a black project angle could have been that new approach.

The answer might lie with Robert Low, the project coordinator, and perhaps the CIA agent that Condon so easily disregarded.

In 1966 when the Secretary of Defense announced the appointment of Dr. Edward U. Condon to head a scientific investigation into UFOs, many serious UFO investigators like J. Allen Hynek, James E. McDonald, Donald Keyhoe, and the Lorenzens had high hopes. This was what they had been waiting for, what the world had been waiting for. Even if the inquiry didn't yield definitive answers it would at least point the way toward further scientific investigation.

But their high hopes were quickly dashed. Early in the project, both Condon and Low made it clear that they regarded UFOs as nonsense and unworthy of serious study. They even made their sentiments public which was in direct violation of the Project's own "rules" which required that no committee member would say anything publicly until the Final Report was issued. Two members were eventually fired and the administrative assistant resigned, citing Low's negative attitude and his lack of interest in sighting reports.

One of the criticisms leveled at Condon was his preoccupation with crackpot cases. He was devoting only a limited amount of his time to the project and it seemed odd that he would spend what little time he had on nut cases while ignoring genuinely credible cases. Unlike James McDonald, who simply filed away the crackpot cases, Condon seemed obsessed with them. He even went so far (literally) as to attend the Congress of Scientific Ufologists in New York City.

Robert Low flatly dismissed UFOs as ridiculous. Instead of taking a serious look at the objects people were seeing, he focused on the witnesses, the implication being that for one reason or another, witnesses could not be trusted or taken seriously. In his notorious memo of August 9, 1966 (see Chapter Two) he wrote that the scientific community would "quickly get the message" if the emphasis was placed on the observers rather than on the UFOs. "One way to do this," he wrote, "would be to stress investigation, not of the physical phenomena, but rather of the people who do the observing--the psychology and sociology of persons and groups who report seeing UFOs." Focusing on the witness instead of on the object had always been a bad habit with UFO investigations. It began .with Ken Arnold when more questions were raised about his character and credibility than about the discs he saw.

But the physical reality of the flying objects was the key to unlocking the UFO mystery. All credible witnesses described odd lighting patterns that belonged to metallic, strangely shaped craft that were capable of fantastic maneuvers and great speed. If Condon and Low had directed their team toward these salient facts how different the Final Report would have been.

It should not have been that difficult for a team of eminent scientists with all the resources available to them to discover the history of the quest for radar-evading aircraft that went back to World War II, to discover the existence of delta winged aircraft, and uncover the ongoing search for ever faster, ever higher, ever more invisible planes.

It is equally mystifying why the Condon Report never mentioned the U-2 as a logical explanation for at least some sightings. The spyplane had flown at an altitude of 60,000 feet at a time when airliners flew at 20,000 feet. At that height, the polished aluminum surface of the U-2 reflected the sun even at night. No wonder witnesses thought it was a UFO. The Blackbird was flying between 80,000 and 100,000 feet at a speed of Mach 3 or even higher. Why weren't these aircraft mentioned?

These are crucial points because the project was funded by taxpayer money and ultimately answerable to the American people, a fact that neither Condon nor Low seemed to appreciate.

Why did Condon and Low take on a project that they didn't believe in? Why did they take taxpayer money to fund a project they considered unworthy? Why didn't they simply decline the air force's offer? Conversely why did the air force pick Edward Condon? No such offer was made to J. Allen Hynek or James McDonald who had equally solid scientific credentials. But Hynek and McDonald differed from Condon (and Low) in one very important respect: they believed there was something to UFOs, and they believed the phenomenon was worthy of scientific study. The two members of the Condon Committee who agreed with this position (David Saunders and Norman Levine) were fired.

Which brings us back to Robert Low. Over the years Low has been roundly criticized for his behavior, his attitude, and especially for his infamous memo. He has been accused of manipulating Edward Condon and sabotaging the Colorado Project. During one argument, Saunders reminded Low "that he had blocked the investigation of one particularly startling UFO case." Levine told Condon how Low "had slammed the door in his face when he brought up the handling by Low of an Edwards Air Force Base case." In her letter of resignation Mary Louise Armstrong, the Administrative Assistant, wrote:

I think there is an almost unanimous lack of confidence in him (Low) as the project coordinator and in his exercise of the power of that position ... Bob's attitude from the beginning has been one of negativism. Bob showed little interest in keeping current on sightings ... Saunders carefully set aside reports on a checkout basis, so that everyone on the committee would have a chance to read them. Bob checked some out, but, to my knowledge, never really read them, and certainly never encouraged the proposed discussions to actually take place. I think he, as project coordinator, should have the initiative to see that this program was carried out .... To me, too much of his time has been spent in worrying about what kinds of "language" should be used in the final report so as to most cleverly avoid having to say anything definitive about the UFO problem. Very little time, on the other hand, has been spent in reviewing the data on which he might base his conclusions.

In his book *UFOs and the National Security State* Richard M. Dolan raises the intriguing possibility that Low was indeed the CIA agent that had been planted in the Colorado Study. During World War II Low had been a combat intelligence officer. After the war, he earned a master's degree at Columbia University in 1948. One year later, the CIA attempted to build an anti-Communist resistance movement in Albania. The agency had taken over this job from the British. The operation was organized by Frank Wisner who later became CIA Operations Chief. One of the men recruited by Wisner was Robert Low. Although the operation failed, it did put the former intelligence officer in touch with Wisner and the newly formed CIA. Given his background, perhaps Low was recruited by the CIA to obfuscate the Colorado Project. His job was made easier by the fact that Condon only spent about half his time directing the project and Low ended up being the de facto director or, as J. Allen Hynek put it, "the actual pilot of the Colorado UFO ship."

The Colorado Project was granted $500,000 by the air force to review all the UFO data up to that time including Project Blue Book cases. The Final Report contained some tantalizing cases that under different leadership would have at least concluded that UFOs, whatever they might be, were real. The possibility of UFOs being secret aircraft might have even been suggested.

When Condon concluded that "further extensive study of UFOs probably cannot be justified in the expectation that science will be advanced thereby," the air force was in effect given permission to stop their UFO investigations. The Final Report came out in 1969, the same year the air force called a press conference and, citing the Condon Report, announced it was closing Project Blue Book. Robert Seamans, Jr., the USAF Acting Secretary, said that continuing Blue Book could no longer "be justified on grounds of national security or in the interest of science."

The conclusions in the Condon Report coupled with the closing of Blue Book effectively cut off any further official, scientific inquiry into the UFO phenomenon as well as any government funding. While UFO investigations would continue, independent groups and individuals that had no official support or financing would conduct them. Ufologists were left to soldier on alone. Even respected scientists like J. Allen Hynek and James McDonald would get no official recognition or help, a situation that continues to this day.

J. Allen Hynek, the scientific advisor for Projects Sign, Grudge and Blue Book who began his career as a confirmed UFO skeptic and then ended up believing that UFOs were real and the government was covering up the truth about them, was deliberately misled by the air force. In investigation after investigation he was stymied, blocked, impeded and lied to. He began to complain publicly of not getting any cooperation from the air force or from any other government agencies. No wonder Hynek became convinced that UFOs were extraterrestrial in origin. But, he was like Condon in one crucial respect; he too seemed unaware of the possibility that UFOs were top secret aircraft. Donald Keyhoe who was also skeptical of UFOs at first came closest to the truth when he announced that the CIA was mainly responsible for UFO cover-ups. But he too thought that UFOs were extraterrestrial and seemed totally unaware of black aircraft.

Either these people were being deliberately misled and fed disinformation or they were in on the secret. From all the evidence, It appears it was the former. It is hard to believe that Hynek would have addressed the United Nations on UFOs in 1978 or founded the organization CUFOS (Center for UFO Studies) if he had known about the black aircraft. Would Donald Keyhoe have written so extensively on flying saucers and UFO conspiracies or headed the NICAP if he had known the truth?

And what of the American public? Over the years, poll after poll showed that the majority of Americans believed there was *something* to UFOs. It wasn't all delusion or fantasy or swamp gas or temperature inversions or weather balloons. They came to believe their military and government was conspiring to keep the truth from them. The American people grew increasingly suspicious and distrustful.

It should not have been that difficult for serious UFO investigators to ferret out the truth. Even before World War II had ended, people were aware that Nazi Germany had developed "wonder weapons" that were more technologically advanced than any other weapons systems of that time. V-I flying bombs and V-2 rockets had devastated London. The Germans had developed the Me262, the first jet and the Me163 Komet, the first and only rocket powered fighter to fly in combat. The logical question was what else did the Germans have in their secret arsenal?

In a secret operation code-named Operation Overcast, U.S. intelligence agents fanned out across Nazi Germany after the war looking for scientists and specialists who had worked on the wonder weapons programs. Dr. Werner von Braun was at the top of the list and his name was already fairly well known. But a "black list" also included scientists like Dr. Walter Miethe who had supposedly experimented with saucer-like aircraft. When the name Overcast was compromised, the code-name was changed to Operation Paperclip. This allowed the U.S. government to grant full immigration status to these Germans.

By 1948 there were 492 German and Austrian scientists, technicians, and specialists working in the U.S. The best known, of course, was von Braun, but there were others like Walter Dornberger who had been von Braun's superior during the war later became Vice President of Bell Aerospace Corporation.

UFO investigators, projects, studies and panels consistently failed to look into the Nazi Germany connection. They paid scant attention to the possible existence of circular wing aircraft or to flying wings. With just a little bit of digging, they should have been able to discover the ongoing search for radar-evading aircraft. It was a known fact that German scientists were living and working in this country. The next logical assumption was that at least some of the advanced German technology was being incorporated into our own post-war projects. Maybe advanced aircraft based on this technology was being built and tested in this country. Perhaps these secret projects were responsible for UFO sightings. But the questions were never asked and the dots were never connected, least of all by the Condon Committee.

While all the UFO hoopla was going on, the stealth program was being developed, tested and perfected. While some people were seeing strange objects and lights in the sky and others were being contacted, transported, and abducted, highly advanced top secret aircraft were taking shape in a program that was more classified than the Manhattan Project.

Black projects were funded secretly. Over the years and decades, billions of taxpayer dollars were spent without public scrutiny and without Congressional oversight or approval. Funding for black aircraft was buried in top secret projects or tacked on to big budgets for white programs. Outrageous sums that were spent on items like toilet seats could have been one conduit for the secret funding. It is likely that at any one time only a handful of top officials knew what was really going on. The USAF and especially the CIA were given a free hand to spend huge amounts of public money on black projects. There was no way of knowing which projects succeeded and which failed, how many test pilots were injured or even killed. There was very little, if any, accountability. Deceit, mendaciousness, cover-ups, and disinformation were all acceptable if they protected the black projects for it was all done in the name of National Security.

Was it all worth it? The answer I think is a qualified yes, because more than any other single factor stealth technology was responsible for the collapse of the Soviet Union and the end of the Cold War.

The Gulf War began on January 17, 1991 when the allies launched an air attack on Iraq. Stealth aircraft led the attack and the Soviet-made Iraqi air defenses simply could not cope. There was no defense against stealth. Even the most advanced Soviet surface-to-air missiles were useless against the F-117 and Team Stealth. After the air attack, the ground war was launched on February 24 and lasted only 100 hours. President Bush I issued a cease-fire three days later.

The real losers in the Gulf War were the Soviets. They realized they could not compete technologically with the U.S. and her allies. If Soviet missiles were useless against an F-117, how could they possibly defend against an even more advanced aircraft like the Aurora? The fact that the UK also apparently had stealth capability also defeated the Soviets. The technology was too far ahead of them. Any effort to catch up would probably bankrupt the country that was already suffering from a seriously depressed economy,

It was in the Gulf War that the previously top secret Stealth aircraft were used extensively in combat for the first time, and in so doing, rendered obsolete air defense systems around the world. By the end of 1991 Mikhail Gorbachev resigned as general secretary of the Communist Party, and in December the Soviet Union officially broke up. Although there were multiple factors involved in the collapse of communism, it is probably no coincidence that these two seminal events -- the Gulf War and the dissolution of the Soviet Union -- bracketed the year 1991.

Area 51, where so many black aircraft were and still are tested, still officially doesn't exist. At an air force press conference about the Roswell incident, Colonel John Haynes, an air force spokesperson, responded to questions about Groom Lake: "There is a facility in Groom Lake, Nevada. Quite frankly, I have no knowledge or expertise in the matter. I understand there are classified things that go on there and that's all I have to say about it. "

In 1997 workers at Area 51 came forward to tell CNN how Groom Lake was used as a secret dumping ground for hazardous wastes. Said one government employee; "The flames and the black, blue and white smoke would form clouds so thick. If the wind were blowing in the direction where you were working, you'd get it all." The stuff was dumped into large open trenches and then set on fire. Workers claimed they developed health problems after breathing in the smoke, but their complaints were ignored and their requests for protective clothing were denied. One worker in particular who developed serious skin lesions was unable to tell his doctor about the burning and the smoke he was breathing because he was sworn to

135

secrecy. After his death, tests showed he had been exposed to dioxins and furons.

Because the government still refuses to admit that Area 51 exists, it has been very difficult for workers to file claims and lawsuits. There is no legal obligation to obey environmental laws, labor laws, or worker rights laws. Attorney Jonathan Turley who represents several injured base workers and the widows of two dead workers claims his clients were threatened with arrest and imprisonment. Said Turley, "Why is the government trying so hard to protect the people who committed these crimes? I've got two dead clients. I've got two people who died. It's not a joke."

When the UFO phenomenon began it was considered rather harmless, but it wasn't long before more sinister elements began to emerge, beginning with the death of Captain Thomas Mantell.

"I've sighted the thing! It looks Metallic--and it's tremendous in size!" With those words Captain Thomas Mantell became the first person to be killed while chasing a UFO. Around noon on January 7, 1948, several people in western Kentucky reported a strange object speeding through the sky. Witnesses said the object was huge, 250 to 300 feet in diameter and looked a bit like "an ice cream cone topped with red."

The object swooped over Godman AFB near Fort Knox where tower operators and the base commander saw it too. At that moment, four Air National Guard F-51 Mustang fighters on a training mission were coming in for a landing. Instead, they were radioed to go have a look. One plane, low on fuel, landed, but the other three Mustangs, led by Captain Mantell, powered up their throttles and climbed toward the object. Mantell was a decorated pilot from World War II, winning both the Silver Star and the Flying Cross.

The object was described by two flyers as metallic and "of tremendous size ... round like a teardrop, and at times almost fluid." After this transmission, two of the F-51 Mustangs quit the chase, but Mantell radioed that he wanted to try for a closer look: "I'm going to 20,000 feet and if I'm no closer then, I'll abandon the chase."
That was his last transmission. A few hours later his body was found among the wreckage of his Mustang near Fork Knox. An air force investigation concluded Captain Mantell had flown too high and

blacked out at 20,000 feet from lack of oxygen. None of the F-51s had been carrying oxygen. And the object had simply been the planet Venus shining in the afternoon sky. Later, they amended their conclusion to Mantell sighting a navy Skyhook balloon. Skeptics still weren't convinced, and rumors began to circulate that Mantell had been blown out of the sky by an alien spacecraft, and the air force was covering it up.

Mantell's son, Thomas Mantell III, six years old at the time, later said the family was not told about his dad's death by the air force. Instead, a neighbor heard about the crash on the radio and came over to tell his mother. "They wanted to sweep it under the rug with as little attention as possible, either because they felt they were at fault or because he was chasing something they didn't. want to talk about." The family still feels resentful and angry toward the air force for not telling them the truth about what really happened.

The wingman who saw the object always maintained they were chasing a "large, golden, iridescent object." The TV show Sightings sent an investigative team to the crash site. The team's digging led to the recovery of over two dozen pieces of aircraft wreckage. Serial numbers on several pieces confirmed it was Mantell's plane. But what was so unusual was that several pieces registered strong levels of radiation even after being buried for over forty years. Also, the wreckage had been buried in an unmarked trench by the same air force that had supposedly done a thorough investigation.

It is unknown how many people involved one way or another with UFOs might have been injured or killed. It is also unknown how many people had their reputations and credibility ruined by making UFO reports. Then there are the cases of the intimidation of UFO witnesses.

On May II, 1950 Paul Trent the farmer in McMinnville, Oregon who took two clear photos of a saucer-like craft described as a huge metallic, very bright, almost silvery disk that flew silently. The case became #46 in the Condon Report where it was described as "moving quite slowly, apparently almost hovering, and it apparently shifted both its position and orientation in a complex way, changing direction and tipping just before it moved away ... " A local newspaper printed the photos Mr. Trent had taken.

After the photos were published, air force personnel in plainclothes as well as FBI agents, the infamous "Men in Black reportedly visited the Trents. MIB (not to be confused with the Tommy Lee Jones and Will Smith movie characters) supposedly belong to a unit of the "Air Force Special Activities Center," and their job is to "visit" witnesses after they report a UFO sighting or experience.

An Air Force sergeant in plainclothes demanded that the reporter who published the photos turn them over. Mr. Trent admitted he had been afraid to let his photos be published for fear of getting "into trouble with the government." Despite the intimidation, the Trent photos became some of the most famous of all credible UFO photos, and are still scrutinized to this day. Even the Condon Committee was impressed: "This is one of the few UFO reports in which all factors investigated, geometric, psychological, and physical appear to be consistent with the assertion that an extraordinary flying object, silvery, metallic, disk-shaped, tens of meters in diameter, and evidently artificial flew within sight of two witnesses."

On the night of December 9, 1965 a strange, firey object landed in a wooded ravine near the small town of Kecksburg in western Pennsylvania. James Romansky, an 18-year old volunteer firefighter at the time, was part of a search and rescue team sent into the woods to look for survivors from what appeared to be an airplane crash. Instead they saw an object that was "like a giant acorn lying on its side. It was made of metal, like bronze, copper, and gold all mixed together." They also noticed odd writing or symbols on the object.

Suddenly two men who appeared to be military approached the firefighters and ordered them to leave the woods, telling them this was now a restricted area. Also, the Pennsylvania State Police had set up a roadblock and cordoned off the area.

Another witness, James Ramsey, saw something else besides the UFO that landed in the woods: "that night ... a flatbed truck came back down off the hill with a Jeep in front of it and a Jeep behind it, and they all had flashing red lights on. There was something loaded on the back of the flatbed covered with a tarpaulin. It was about the same size as the object I saw."

A group of military people came to a local farmhouse belonging to the Hays family shortly after the fireball landed in the woods behind their house. John Hays, ten years old then, recalled them taking over the house and setting up a kind of command center. His parents were told to send the kids to bed.

The official version of the Kecksburg UFO concluded that a thorough investigation had been done and "nothing whatsoever" was found in the woods. The witnesses who had come forward were made to look foolish or like fakers or liars. Other witnesses later said they didn't want to speak up because they were afraid of the state police or didn't want to get in trouble with the army. Consequently, the small town became divided between the Believers and the Non-believers, a division that persists to this day.

Just how real are the MIB? Do they really go around harassing and intimidating UFO witnesses? Do they infiltrate amateur UFO groups and even scientific projects in order to discredit them? Do they use ridicule and intimidation to get people to deny what they saw or to change their stories?

According to the Sci Fi channel's special on Rendlesham Forest, the MIB belong to the OSI (Office of Special Investigation). They are a police force in civilian clothes who use a combination of methods including intimidation, chemicals, hypnosis and trickery to get witnesses to deny their experiences and sign false documents

In the Rendlesham Forest case, according to the Sci Fi channel, military witnesses were subjected to "intense interrogations." Colonel Charles Halt believes some of them are still suffering from the effects of the techniques that were used on them, especially Larry Warren who was given an injection and "worked on."

MIB are also thought to work behind the scenes by spreading disinformation. Mind control techniques are used to get people to tell such bizarre and crazy stories that no one believes them, and their credibility and perhaps even their reputations are destroyed. Larry Warren was purportedly taken to a secured facility and made to sign false documents. He was shown films of UFOs and told "bullets are cheap." Warren later claimed that he suffered from nightmares and post traumatic stress.

In 1994 Jim Penniston, another Rendlesham witness,

underwent hypnotic regression by a psychiatrist to help eliminate his nightmares. While under hypnosis, he remembered being given sodium penethol and questioned by two mysterious men. He remembered being debriefed by DS8 which he later learned was Defense Secretary at Eight, responsible for leading UFO investigations.

An interesting dramatized version of MIB in action is the film *The Disappearance of Flight* 412 starring Glenn Ford, David Soul, and Bradford Dillman. In this film, a group of air force personnel is interrogated and worked on by special agents after they witness a UFO encounter. Hero Glenn Ford encounters nothing but frustration and obfuscation and threats as he tries to learn the truth about what happened to his men and what they saw.

The truth about UFOs is that they really do exist. They are not the products of fantasy, delusion, or mass hysteria. But they are not from outer space, **they come from planet Earth**, and they are made by us. They are highly advanced, top secret black aircraft that look as alien as anything from outer space. If the American people had been told about black projects from the very beginning, there would have been no UFO Phenomenon, no contactees, transportees or abductees.

One of the most tragic legacies of black projects is the wake of damaged and ruined reputations it left behind. From the very beginning, UFO investigators focused on the witnesses rather than the objects. Kenneth Arnold, "the man who saw the Men from Mars," found himself defending not only what he had seen, but his reputation as well.

The insidious practice of trashing witnesses and character assassination took precedence over scientific analysis of the actual objects. Given the level of intimidation, it is amazing in retrospect that so many credible witnesses had the temerity to come forward.

It is time for our government to come clean about the black projects that gave rise to the UFO myth. The intense secrecy surrounding top secret aircraft must end. The Cold War is over and the *raison d'etre* for them is gone. The America people deserve no less because, after all, we paid for them.

11. THOSE DEADLY DRONES

Technology is advancing so fast that it is difficult to keep up with it. Such is the case with unmanned aircraft. What started out sixty years ago as a backyard hobby of remote contolled model airplanes has turned into an enormous unbelievable array of objects flying in the sky. Will it every slow down or come to a place where rapid advancement is no longer feasible?

An **unmanned aerial vehicle (UAV)**, best known as a **drone**, is an aircraft without a human pilot. Its flight is controlled either autonomously by computers in the vehicle, or under the remote control of a pilot on the ground or in another vehicle. There are a wide variety of UAV shapes, sizes, configurations, and characteristics. UAVs were originally hobby type remote controlled model planes that became sophisticated weapons of destruction. UAVs are used mostly for military operations, but are also being employed for non-military uses such as policing, firefighting, security surveillance, search and rescue. They can check pipelines for leaks. They are preferred for military missions that have odds too dangerous for human engagement. The United States had not yet arrived at the stage where we recruit suicide or "kamikaze" pilots, so we started unmanned flights in a big way.

To distinguish UAVs from *missiles*, a UAV is defined as a "powered, aerial vehicle that does not carry a human operator, uses aerodynamic forces to provide vehicle lift, can fly autonomously or be piloted remotely, can be expendable or recoverable, and can carry a lethal or nonlethal payload".

Missiles are not considered UAVs, because, like many other guided missiles, the vehicle itself is a weapon that is not reused, even though it is also unmanned and in some cases remotely guided.

The United States has two separate UAV programs, one run by the military and the other by the Central Intelligence Agency. The military's UAV program is overt, meaning it is recognized by the public and therefore only operates where US troops are stationed. The CIA's program is covert. Missions performed by the CIA's UAV program do not always occur where US troops are stationed.

The CIA got into the program as a result of the September 11 attacks on our homeland. Today, the UAVs are definitely used in Afghanistan, Pakistan, Yemen, and Somalia. Under the codename "Eagle Program" the CIA operates foreign surveillance programs.

Drones and other Unmanned flying machines.

Some of the terms in use to identify the various drones are UAS -an unmanned aircraft system, UAV-an unmanned air vehicle system, and UCAV – an unmanned combat air vehicle. The term UAV will be used in this report. There is also RPA used by the U. S. Air Force for remotely piloted aircraft.

UAVs typically fall into one of six functional categories.

1. Target and decoy – providing ground and aerial gunnery a target that simulates an enemy aircraft or missile
2. Reconnaissance – providing battlefield intelligence
3. Combat – providing attack capability for high-risk missions
4. Logistics– UAVs specifically designed for cargo and logistics operation
5. Research and development – used to further develop UAV technologies to be integrated into field deployed UAV aircraft
6. Civil and Commercial UAVs– UAVs specifically designed for civil and commercial applications

UAVs can also be categorized by range of operation and operational altitude. These have different names. The first number after the name indicates the range in feet and the second number the altitude of maximum performance. Hand-held 2,000 – 600, Close 5,000- 1,500, NATO 10,00 – 3,000, Tactical 18,000 – 5,500, MALE 30,000 – 9,000, HALE 30,000 – 9,000.

MALE stands for medium altitude, long endurance and HALE stands for high altitude, long endurance. There are also categories of Hypersonic, Orbital, CISLunar and CACGS Computer.

Brief History

The "Areal Target" by A.M. Low created in 1916 was the first publized attempt at creating an umanned aerial vehicle. A year earlier electrical engineer Nicola Tesla had written an ariticle on the possibility of unmanned aerial combat vehicles.

During and shortly after World War I there was a flurry of technical advances in remote controlled aircraft. These were used in World War II to train anti-aircraft gunners and fly attack missions. Germany used UAVs during the war with some success. Shortly after the war, jet engines were installed in these experimental craft. The Beechcraft company was commissioned in 1955 by the U.S. Navy to develop their Model 1001.

Production of UAVs increased dramatically during the Vietnam Era. Previous to that time the United States was using high aaltitude spy planes. This became dangerous since our enemy at that time, the Soviet Union, had developed missiles and Francis Gary Powers piloting the secret U-2 plane was shot down. The U. S. Military Commanders immediately launched a UAV program code-named "Red Wagon." So during the Tonkin Gulf incident starting on August 2, 1964 used the highly classified UAVs with immediate success. Some of the UAVs were shot down and they were displayed to the world by the Chinese. The United States response to these photos was "no response."

It was the testmony in Februrary 1973 for the United States Housee Committee on Appropriates that the U.S. Military officially confirmed that they had been utilizing UAVs in Vietnam. At that time over 5,000 airman had been killed and 1,000 more wee missing in action or, at that time, possible captured. The U.S. Air Force had flown close to 3,500 UAV missions during the war. On these missions "only" about 550 were lost. How many American pilot lives were saved can't be calculated, but the number must be in the hundreds.

In the same year mentioned above, 1973, Soviet surface to air missiles began shooting down Israeli jets during their war with Egypt and Syria. Israeli scientists began the development of the first modern UAVs and shared this knowledge with the United States. The photo images obtained by the UAVs helped Israel to neutralize the Syrian air defenses at the start of their skirmish with Lebanon in 1982. Israel devcloped tail-less stealth technology based on three-dimensional thrust vectoring flight steering in 1987. This was shared with the United States and this should leave no doubts about the why of our country's association with Isreal. The U.S. Department of Defense has purchased many UAVs from a joint Amercian corporation working with the Israeli company Mazlat.

One of the first UAVs was the Pioneer, which was used to identify the location of artillery and boats. This was followed by the Predator. Between 2008 and 2011, the U.S. Air Force has used 5,331 UAVs which was twice the number of manned planes employed at that time.

A drone: General Atomics MQ-1 Predator

The Predator has a Hellfire Missile that can terminate any target it locates. The first Predators captured Osama bin Laden on film, but the technology at that time was such it couldn't do anything about it except send back images. That would not be true today since the predator can designate a robot to attack. In 2005 to 2006 Predators carried out 2,073 missions and participated in 242 raids. The number of missions for the Predator after that is highly classified.

The Predaotor is remotely controlled by satellites by pilots located 7,500 miles away. The new UAV, the Global Hawk, operates virtually autonomously. The user merely hits the button for 'take off' and for 'land', while the UAV gets directions via GPS and reports back to the station command post. Global Hawks can fly three thousand miles and map an area the size of Montana in a matter of minutes.

The big development in UAVs is in miniaturization. The small units can be launched by hand and maneuvered up and down a street or an alley. They can float around for a week. Terrorists usually stay in under cover during the day since this small UAV known as The Raven can spot them while being undetected. Once the Raven gets its picture, it's "good night Irene" for the insurgent.

Assassination and Legality

A confidential Justice Department memo concludes that the U.S. government can order the killing of American citizens if they are believed to be "senior operational leaders" of al-Qaeda or "an associated force. This is legal even if there is no intelligence indicating they are engaged in an active plot to attack the U.S. The secrecy surrounding such strikes is fast emerging as a central issue in many closed-door meetings. Attorney General Eric Holder endorsed the constitutionality of targeted killing of Americans with justification that the government determines the target as "an imminent threat of violent attack." This is justified under the concept of the right of self defense. This also involves a broader concept of imminence than actual intelligence about any ongoing plot against our country.

The present concept of using UAVs in targeted killings considers whether an attempted capture of a suspect would pose an "undue risk" to U.S. personnel involved in such an operation. If so, U.S. officials could determine that the capture operation of the targeted American would not be feasible, making it lawful for the U.S. government to order a killing instead.

The present government concept of this matter includes a more extensive discussion of why targeted strikes against Americans does not violate constitutional protections afforded American citizens as well as a U.S. law that criminalizes the killing of U.S. nationals overseas. It also includes why such targeted killings would not be a war crime or violate a U.S. executive order banning assassinations. The reasoning behind this is that "A lawful killing in self-defense is not an assassination."

An Homeland Security document states "In the Department's view, a lethal operation conducted against a U.S. citizen whose conduct poses an imminent threat of violent attack against the United States would be a legitimate act of national self-defense that would not violate the assassination ban. Similarly, the use of lethal force, consistent with the laws of war, against an individual who is a legitimate military target would be lawful and would not violate the assassination ban."

New Military Roles

The military role of unmanned aircraft systems is growing at unprecedented rates. In 2005, tactical- and theater-level unmanned aircraft alone had flown over 100,000 flight hours. Rapid advances in technology are enabling more and more capability to be placed on smaller airframes which is spurring a large increase in the number of Small Unmanned Aircraft Systems (SUAS) being deployed on the battlefield. The use of SUAS in combat is so new that no formal reporting procedures have been established to track SUAS flight hours. As the capabilities grow for all types of UAS, nations continue to subsidize their research and development leading to further advances enabling them to perform a multitude of missions. UAS no longer only perform intelligence, surveillance, and reconnaissance missions, although this still remains their predominant type. Their roles have expanded to areas including attack as well as strike missions, suppression and destruction of enemy air defense, network mode or communications relay. These UAS range in cost from a few thousand dollars to tens of millions of dollars, with aircraft ranging from less than one pound to over 40,000 pounds.

When the Obama administration announced in December 2009 the deployment of 30,000 new troops in Afghanistan, there was already an increase of attacks by unmanned Predator UAVs against Taliban and al-Qaeda militants in Afghanistan and the tribal areas of Pakistan. There were 43 attacks between January and October 2009. Between 2006 and 2009, UAV-launched missiles allegedly had killed between 750 and 1,000 people in Pakistan, according to the report. Of these, about 20 people were said to be leaders of al-Qaeda, Taliban, and associated groups. Overall, 66% to 68% of the people killed were militants, and 31% to 33% were civilians. U.S. officials disputed the percentage for civilians. The U.S. Air Force has recently begun referring at least to larger UAS like Predator, Reaper, and Global Hawk as Remotely Piloted Aircraft (RPA), to highlight the fact that these systems are always controlled by a human operator at some location.

146

Artificial intelligence is advancing to the point where the aircraft are easily capable of taking off, landing, and flying themselves. Then they simply have to be instructed as to their mission. The military distinguishes between piloted and supervised systems. A piloted system would be run by an operater while a supervised system would be observed as the operation was underway. As of 2013, few fully autonomous systems have been constructed, but this is more a matter of convenience and technical implementation rather than any fundamental barrier.

Other UAV Roles

Remote sensing operations include electromagnetic spectrum analysis, These include visual spectrum and infrared cameras as well as radar systems and biological sensors capable of detecting airborne microorganisms and other biological factors. Thus, we can send in a UAV to authenticate the use of biochemical weapons of mass destruction.

Aerial surveillance of large areas is made possible with low cost UAV systems. Surveillance applications include: livestock monitoring, wildfire mapping, pipeline security, home security, road patrol and anti-piracy. The trend for use of UAV technology in commercial aerial surveillance is expanding rapidly with increased development of automated object detection approaches.

UAVs are increasingly used for *domestic police work* in Canada and the United States. This is a bone of contention that concerns individual liberty and security.

UAVs can be used to perform *geophysical surveys,* in particular geomagnetic surveys where the processed measurements of the differential Earth's magnetic field strength are used to calculate the nature of the underlying magnetic rock structure. A knowledge of the underlying rock structure helps trained geophysicists to predict the location of mineral deposits. The production side of oil and gas exploration and production entails the monitoring of the integrity of oil and gas pipelines and related installations. For above-ground pipelines, this monitoring activity could be performed using digital cameras mounted on one, or more, UAVs.

147

UAVs can *transport goods* using various means based on the configuration of the UAV itself. Most payloads are stored in an internal payload bay somewhere in the airframe. This is an area where much more experimentation is demanded.

Super storms can be entered by UAVs where it might be too dangerous for manned aircraft. Australia has a 35 pound UAV that can fly into a hurricane and record important data as well as the standard temperature and pressure. These can also be used in severe climates such as found over the polar caps.

Some Success ? Stories

A former Pakistani ambassador to the United States claimed that ten civilians died for every military deaths by UAV attacks. These attacks were turning Pakistani opinion against the United States, and that 35 or 40 such attacks only killed 8 or 9 top al-Qaeda operatives.

By October 2009, the CIA claimed to have killed more than half of the 20 most wanted al-Qaeda terrorist suspects in targeted killings using UAVs. By May 2010, counter-terrorism officials said that UAV strikes in the Pakistani tribal areas had killed more than 500 militants since 2008, and no more than 30 (5%) nearby civilians—mainly family members who lived and traveled with the targets. UAVs linger overhead after a strike, in some cases for hours, to enable the CIA to count the bodies and attempt to determine which, if any, are civilians. A Pakistani intelligence officer gave a higher estimate of civilian casualties, saying 20% of total deaths were civilians or non-combatants.

After more than 30 UAV-based strikes hit civilian homes in Afghanistan in 2012, President Hamid Karzai demanded that such attacks end, but the practice continues in areas of Pakistan, Yemen, and Somalia that are not in war zones.

In 2012, the U.S. Air Force trained more UAV pilots than ordinary jet fighter pilots for the first time. The new technology seems to have no limits.

An unofficial report in 2009 stated that an Air National Guard pilot in an F-16 saw flashing lights underneath him while flying over

Afghanistan at twenty-three thousand feet and thought he was under fire from insurgents. Without getting required permission from his commanders, he dropped a 500-pound bomb on the lights. They instead turned out to be troops from Canada on a night training mission. This is important because if that plane had been a UAV, it would have had more time to make its decision. Since the UAV has long- range cameras, it would have been able to get a visual on the potential target and discover that they were friendly. In addition, it also would not feel fear or adrenaline that could potentially affect the decision of a person who fears for their life.

In February 2013, U.S. senator Lindsey Graham stated that 4,756 people have been killed by U.S. UAVs. The accuracy and relevance of that statement requires further discussion.

A Final Note: There are unfounded rumors at this time that the U.S. has developed a miniature UAV the size of a mosquito that can dive in, hit a person, get a blood sample and move out.

TIMELINE
February 27, 1910
Clarence "Kelly" Johnson is born. He will head the Skunk Works, Lockheed's secret aviation development department.

1916
A.M. Low develops a workable Uas named Aerial Target

1933
In Germany, Reimar and Walter Horten test an all-wing sailplane. Adolph Hitler becomes Reich Chancellor.

September 1939
Hitler invades Poland, and World War II begins.

June 18, 1941
Germany invades the Soviet Union. Prior to the invasion, high altitude secret reconnaissance flights were made over Soviet territory with specially modified aircraft.

December 7, 1941
The Japanese attack Pearl Harbor, and the United States enters the war.

February 1942
The Battle of Los Angeles.

December 1942
The Northrop N-9M, a flying wing, is flown for the first time.

1944-1945
Nazi Germany builds wonder weapons including the first rockets, guided missiles, and jets.

1945
World War II ends, and almost immediately the Cold War begins. The U.S. and Soviet Union scramble to capture the wonder weapons and the scientists and technicians who developed them. Both countries carry out secret ops.

June 1946
The Northrop XB-35, a flying wing bomber, flies for the first time.

Northrop's flying wing bomber

June 24, 1947

Pilot Kenneth Arnold sees "nine very bright objects" flying over the Cascade Mountains. An enterprising reporter dubs the objects "flying saucers" and a phenomenon is born.

July 1947
The nation's armed forces are unified, including the new air force. James Forrestal is the first Secretary of Defense. The National Security Council and the Central Intelligence Agency are created.

July 2, 1947
Rancher William "Mac" Brazel recovers the remains of what some think is a flying saucer. This will become the Roswell Incident.

September 1947
Project Sign begins an investigation of flying saucers. In 1949 it will become Project Grudge, and from 1951 until 1969 it will be Project Blue Book. J. Allen Hynek, an astrophysicist at OSU, will be the scientific consultant for all three

October 1947
Major Chuck Yeager breaks the sound barrier in his Bell X-I rocket plane.

January 7, 1948
Captain Thomas Mantell is killed while chasing a UFO. Coming so soon after Roswell, his death fuels rumors and conspiracy theories.

January 22, 1948
Project Saucer is begun by the Air Material Command's Technical Intelligence Division at Wright Field in Dayton, Ohio. J. Allen Hynek works under contract as an independent investigator.

April 27, 1949
Project Saucer's preliminary studies are released to the press. AMC's Technical Intelligence Division looked into 240 domestic and 30 foreign incidents and concludes "exhaustive investigations have turned up no alarming probabilities ... the question marks in

Project Saucer are not dangerous ones."

May 11, 1950
Paul Trent of McMinnville, Oregon takes the first credible photos of
a UFO.

1954
Lockheed's Kelly Johnson helps design the F-I04 Starfighter that can
fly at Mach 2. The U-2 spy plane is also developed at the Skunk
Works, and it becomes the first aircraft to fly above 60,000 feet..

1955
Beechcraft builds the UAV known as Model 1001

June 2,1957
Captain Joseph Kitttinger makes the first Manhigh flight as part of a
secret high altitude balloon experiment. Riding in a pressurized
gondola, he reaches an altitude of 97,000 feet, the edge of space. In
August Manhigh II reaches the program's highest altitude -- 101,516
feet.

October 4, 1957
The Soviet Union launches its sputnik satellite, and the space race
begins.

June 8,1959
The X-I5 is dropped from its B-52 mother ship. The rocket plane
flies at hypersonic speed (Mach 6), and its ceiling takes it to the
edge of space. The X-I5 pilots are awarded astronaut wings. One of
them is Neil Armstrong.

1960
Project Ozma, the first attempt to detect radio signals from an
extraterrestrial civilization, is directed by Dr. Frank Drake at the
National Radio Observatory in Green Bank, West Virginia.

May 1, 1960

A U-2 spyplane is shot down over Sverdlovsk, USSR, the first time a missile is used to bring down an aircraft. The U.S. decides it needs an invisible plane because "You can't hit what you can't see,"

1962 -- 1964
The A-11, a CIA reconnaissance aircraft, flies for the first time in April 1962. A similar version, the A-12, is developed but with a lower radar signature. In 1963 an interceptor version, the YF-12A, is developed. The top-secret SR-71 Blackbird is developed from the YF-12A, and flies for the first time on December 22, 1964.

1964
U.S. uses UAVs in the Gulf of Tonkin, (Vietnam Conflict)

September 1964
Lieutenant Robert Jacobs films an Atlas missile test flight at Vandenberg AFB. When the footage is shown, a UFO can be seen flying around the missile and emitting flashes of light.

January 1966
The first SR-71 Blackbird enters service. It is delivered to Beale AFB in California.

March 20, 1966
UFOs are spotted near Hillsdale Michigan. J. Allen Hynek, the scientific consultant to Project Blue Book, gives his famous "swamp gas" explanation.

1969
A bad year for ufologists. The Condon Committee's *Scientific Study of Unidentified Flying Objects* is published, and Project Blue Book is terminated.

1972 Dr. J. Allen Hynek publishes *The UFO Experience* in which he classifies "close encounters," and pleads for a more scientific approach to the study of UFOs

October, 1973
Jeff Greenhaw, the police chief of Falkville, Alabama, sees a metallic suited creature in the middle of the road while investigating a report of a UFO landing.

October 18, 1973
The Coyne Helicopter Incident.

1975
Denys Overholser at the Skunk Works begins working on stealth or low observable technologies. He creates a software program that refines and simplifies the calculations of Pyotr Ufimstev, the Russian physicist and father of stealth.

December 1, 1977
The first flight of the Have Blue prototype takes place at Groom Lake amid tight security. Have Blue will evolve into the F-117 A.

December 27-29, 1980
The Bentwaters/Woodbridge/Rendlesham Forest Incident in the UK is likened to Roswell in the U.S.

1982
The F-117 stealth fighter is deployed. With air refueling, its range is unlimited.

February 1982
Northrop's Tacit Blue makes its first flight. Although this test vehicle never goes into production, it yields valuable data on the use of stealth aircraft for reconnaissance.

May 5, 1984
The "Fast Walker" UFO is tracked by a USDSP satellite.

February, 1985

In its procurement budget the Pentagon requests the" Aurora" project be placed next to the SR-71. In 1986 $80 million is spent, and in 1987 $2.3 billion is spent. The code name Aurora then disappears.

1987
Israel develops an advanced system of UAVs and shares this technology with the United States

1988
A *New York Times* article reports that a reconnaissance, hypersonic aircraft named Aurora has been developed to replace the SR-71 Blackbird. Aurora may have first flown in 1985, and its speed is reportedly in the Mach 6-8 range. Aurora is above top-secret.

1989.
Chris Gibson, a trained aircraft observer working as an oil-drilling engineer in the North Sea, sees a mysterious triangular aircraft flying between two F-111 fighter-bombers and a Hercules tanker. Some believe this is the first reported sighting of the Aurora.

July 17, 1989
The B-2 stealth bomber flies for the first time. Its wingspan is larger than half of a football field, but its radar cross section is the size of a bird.

1990
The Pentagon announces that after 28 years it is retiring the SR-71 Blackbird, and will rely on satellites for surveillance. But according to rumors the replacement is actually a hypersonic, top secret spyplane.

1991
A series of "skyquakes" are detected in southern California by seismologists who estimate the aircraft are flying at Mach 3-4.

1991

The F-117 stealth fighter is used in the Gulf War. The top-secret aircraft has now entered the white world.

1992
Steven Douglas of Amarillo, Texas photographs the "donuts-on-on-a-rope" contrail believed to be left behind by the Aurora's PDE engines.

1994
Boeing teams with Lockheed Martin to develop DarkStar, a stealthy UAY for the U.S. Department of Defense Tier III Minus program.

1999
The B-2 stealth bomber is used in combat for the first time in Kosovo.

.

ADDENDUM

Airships
  Arguably the first UFO wave, it began in 1896 and lasted until April 1897. On the evening of November 17, 1896 a light described as an "electric lamp propelled by some mysterious force" flew over Sacramento, California. The *San Francisco Chronicle* said it was all a hoax and put it down to a hot air balloon with an electric light attached to it. The airship was spotted again on November 20 in Sacramento and in San Francisco on November 21.One witness who saw it on November 26 described a 100-foot long body made of aluminum with a triangular tail.

  Others claimed they saw the airship land in various places in order for the human-like crew to make repairs. Between February and April 1897 witnesses reported the airship now had a "big searchlight" on it. In Decatur, Michigan, people reported a bright light with a dark shape behind it. They also said they heard voices and crackling sounds. On April 17, 1897 witnesses in Aurora, Texas claimed the airship flew low over the town, zipped over the town square and then headed north. It struck a windmill on a nearby farm and was destroyed. When people ran to the scene they found the dead pilot among the wreckage. The airship sightings, whether fact or fantasy or a combination of both, occurred right before the Spanish American War (1898), and there was speculation at the time that it was a secret weapon for bombing Cuba.

ALH84001

A meteorite discovered in the Allan Hills of Antarctica in 1984. In 1994 it was classified as belonging to the SNC meteorite group which means it originated on Mars. It is 4.5 billion years old, and landed in Antarctica some 13,000 years ago. The patterns of the linked minerals in the meteorite's cracks resemble patterns produced in rock by microorganisms on Earth. Also, the rock contained organic compounds found in coal and petroleum as well as worm like segments similar to fossilized bacteria on Earth. Some concluded this indicated primitive microbial life on Mars, but others thought it was just mineral formations.

Area 51

A top secret U.S. government installation in Nevada where classified black projects are developed and tested. The site is part of Groom Lake AFB.

Arecibo

A 1,000-foot radio telescope in Puerto Rico used to search for radio transmissions from alien civilizations. The observatory was also used to broadcast a message from Earth. This message was beamed to M 13, a distant globular star cluster. Arecibo is part of NASA's High Resolution Microwave Study (HRMS). The Serendip team at UC Berkeley also uses the dish to search for alien signals.

AST 340

A joint American and British project to explore the possibility of constructing nuclear powered aircraft.

Astrobiology

The study of the condition of life in the Universe. Life on Earth might be only one example of how life works; there may be other models. Astrobiology also looks at entire planets as ecosystems.

Avrocar

A UFO built by the USAF and U.S. Army during the 1950s and 1960s. Built in Toronto, Canada, it was circular in shape and

157

looked like a flying saucer. Because of a stability problem, it supposedly never worked to its estimated potential.

AX-17

A stealth attack plane that may or may not exist. It is believed to be a replacement for the F-111 Fighter-Bomber. It's a swing-wing aircraft, and is stealthiest in its maximum sweep.

Back-Engineering (reverse engineering)

The theory that black aircraft technology is based upon captured UFOs presumably held at Area 51. The Aurora spy plane is rumored to be one result. Some believe the U.S. is in communication with UFOs and the ETs are giving us powerful technology including advanced forms of weaponry and propulsion.

Lord Hill-Norton, the Former Admiral of the Fleet, believes an exchange of technology has taken place between the U.S. and the UFO occupants. American technological advances were made possible by information given to us by the aliens.

Bandits

Pilots who fly the black aircraft. Their missions, like their aircraft, are top secret. They wear pressurized suits like the astronauts.

Battle of Los Angeles

February 25, 1942, also known as the Los Angeles Air Alarm. Radar picked up unidentified aircraft, including one very large plane approaching Los Angeles. Because officials thought it was a Japanese invasion, the city was blacked out. The huge ship hung over Culver City and Santa Monica for nearly an hour while army anti-aircraft batteries fired some 1,430 rounds of 12-pound high explosive shells at it to no effect. The event was witnessed by an estimated one million people. A Navy Department investigation concluded that it was wartime jitters and the gunners had been firing at empty sky.

Beta B-120

A titanium alloy developed for the SR-71 Blackbird that can absorb and deflect radar waves.

Black Mailbox

A mailbox along Nevada's "ET Highway" which is the closest physical landmark to Area 51. It is now said to be painted white.

Black Manta (TR 3-A)

A stealthy, triangular shaped, reconnaissance aircraft designed to relay information on targets to attack aircraft like the F-117. It was supposedly used for this purpose in the Gulf War. The aircraft may use a laser to illuminate a target in order to guide bombs. A Black Manta purportedly crashed at Area **51 in** November 1987.

Brilliant Buzzard

A large, delta wing aircraft secretly built by Boeing in the 1980s. It is reportedly over 200 feet long with a cruising speed of Mach 3. It is believed to be light-colored on the top and bottom, with dark leading and trailing edges. It may be a launching vehicle for a smaller, hypersonic aerospace plane. Because its name includes "brilliant" it is believed to be part of the Strategic Defense Initiative (SOl).

Brilliant Pebble

The code-name for SDI. Any project with "brilliant" in its name is supposedly part of SDI.

Buron

Russia's version of the space shuttle. Its name translates to "blizzard." It reportedly flew one time and unmanned in 1988.

Carter, Jimmy

The first major political figure to publicly admit seeing a UFO. The sighting occurred in January 1967 when he was governor of Georgia. Governor Carter filed two formal reports describing his sighting that he said was a red and green glowing orb that streaked across the southwestern sky, only to vanish ten minutes later.

Cheney, Richard

As Defense Secretary under President Bush I, he supposedly designated the Aurora spy plane as a "waived program," meaning that it is so secret only the chairmen and ranking members of the House and Senate military committees know of its existence.

Churchill, Winston

In 1952 Churchill asked Lord Cherwell, his air minister, to investigate UFOs. Churchill reportedly asked: "What does all this stuff about flying saucers amount to? What can it mean? What is the truth?" Twelve days later he was told there was no evidence that

UFOs existed.
CIA (Central Intelligence Agency)
The CIA is an independent executive branch of the U.S. government established by the National Security Act of 1947. The agency replaced the Office of Strategic Services (OSS) which had operated during 1942-1945 and was the first U.S. intelligence agency. In 1949 the CIA was given special powers which allows the director to spend agency funds without accounting for them.

The size of the staff is secret and employees are exempt from civil service procedures. The agency's main responsibilities are to gather intelligence abroad and to perform covert operations. The CIA was mainly responsible for the UFO phenomenon; directing, funding, and hiding classified tests and projects.
Circular-Wing Aircraft
A type of aircraft believed to have been built first in Nazi Germany. In 1939 Heinrich Focke had patented a saucer-like aircraft. Another aircraft was built by Messerschmitt, the AS 6. Dr. Walter Miethe was reportedly the director of the German Circular Wing program. The American V -17 or "flying flapjack" was designed for the U.S. Navy in 1942. The main purpose of these aircraft was to evade radar detection. The circular saucer shape was thought to be ideal for reducing an aircraft's RCS.
Cold War
The long, dangerous standoff between the United States and the former Soviet Union. Although the two countries had been allies during World War II, they quickly became bitter rivals after the war's end. The U.S. and her allies formed NATO while the Soviet Union and her allies became the Warsaw Pact. For nearly 50 years the Cold War waxed and waned with intermittent crises that threatened to erupt into a third world war with thermonuclear weapons. The Cold War finally ended with the break-up of the Soviet Union on December 26, 1991. To be fully understood the UFO phenomenon must be seen in the context of the Cold War.
Comanche
The RAH-66 Comanche is the world's most advanced helicopter. It incorporates stealth technology and can reach a speed of 175 knots with a cruising speed of 165 knots. It can also execute snap

turns in less than five seconds and fly sideways or backwards at 70mph. It can climb 1,418 feet per minute. The Comanche is equipped with 14 Hellfire anti-tank missiles that are programmed to control their own flight to the targets as well as 56 rockets or 28 Stinger missiles. The helicopter also has a fully retractable missile armament system that allows it to hide its missiles and rockets in to maintain its low RCS. The Comanche may be part of Team Stealth.

COMETA

"Committee for In-depth Studies." A study on UFOs completed in France by an independent group supposedly under the auspices of France's Advanced Studies for National Defense (IHEDN). The report, published in 1999, concluded UFOs are real and are extraterrestrial in origin. The committee considered the hypothesis of secret weapons, but dismissed it. IHEDN denies any official involvement in the study.

Copper Canyon

A classified program begun by the Defense Advanced Research Projects Agency (DARPA) in June 1983. The purpose of the program was to develop air-breathing, hypersonic and single stage, stage-to-orbit (SSTO) aircraft. Concepts generated by the multi million dollar program included high speed cruise missiles, two stage to orbit satellite launchers, high-Mach bombers and reconnaissance aircraft and possibly hypersonic aircraft. The NASP (National Aerospace Plane) is also believed to have come out of the program.

DarkStar

A stealthy unmanned reconnaissance aircraft designed and built in 1994 by Lockheed Martin Skunk Works and Boeing Defense and Space Group for the U.S. Department of Defense Tier III Minus program. DarkStar looks like a flying saucer with wings. Its wing span is 69 feet and its length is 15 feet with a ceiling of over 45,000 feet. It flies at subsonic speeds. Its missions last more than eight hours, providing data from anywhere inside enemy territory, day or night, in all kinds of weather. The program was terminated in 1999 when it was replaced by more advanced UA V s like the Predator.

DARPA

The Defense Advanced Research Projects Agency. From the agency's home page: "is the central research and development organization for the Department of Defense (DoD), It manages and directs selected basic and applied research and development projects for DoD, and pursues research and technology where risk and payoff are both very high and where success may provide dramatic advances for traditional military roles and missions."

"Deuce"

The Convair F- 102 was the first supersonic warplane that had a triangle-shaped wing. It was armed with rockets including a Falcon missile with an atomic warhead. Its maximum speed was 825 mph with a 54,000-foot ceiling. It was the air force's main interceptor in the 1950s.

DFS (Deutsches Forschungsinstitut fur Segelfug)

At the end of World War II, the Russians discovered the German Research Institute for Sailplanes. The Russians found four almost completed, high altitude, supersonic rocket planes. The DFS aircraft, along with most of their specialists, were sent back to the Soviet Union. Concern grew in the US intelligence community that some UFOs might actually be secret Russian aircraft (based on secret German technology) that had penetrated American airspace undetected.

Disinformation

The spreading of false and/or misleading information. In the case of UFOs it is alleged that CIA agents penetrated amateur UFO groups in order to spread and promote the idea that UFOs were extraterrestrial spacecraft and to make such groups appear loony. In addition, the CIA was thought to have manipulated the Press into printing false UFO stories. Some also believe CIA agents penetrated serious UFO projects like the University of Colorado Project in order to steer investigators away from highly classified black projects.

DNA (Deoxyrionucleic acid)

The molecule upon which all life on Earth is based. The "double helix" is made up of adenine, cytosine, guanine, and thymine. Is DNA the only molecule in the universe that can form the basis of life or are there other life-giving molecules? On Earth, DNA

is the only molecule capable of replication and evolution.
Domed Discs

A type of UFO that was reported during the late 1940s and early 1950s. Described as flat-based, cupola-crowned discoids, they were probably the forerunners of stealth aircraft.
DM-1

A small triangular aircraft designed by Alexander Lippisch and a group of students from the Universities of Darmstadt and Munich in Germany during World War II. The DM-1 was to be carried on a larger aircraft to a height of 26,000 feet before it was released. In 1944. the aircraft reached a speed of 500 mph, but it could also fly as low as 40 mph. After the war, it was sent back to the U. S. where it was tested and studied and then used in later delta-wing aircraft projects.
Drake, Frank

While a radio astronomer at the National Radio Astronomy Observatory in Green Bank, West Virginia, he carried out the first attempt (1960) to detect radio signals from alien civilizations. He also formulated the Drake Equation; a method for estimating the number of advanced technical civilizations that may exist in our Milky Way Galaxy. He now serves on the Board of Trustees of the SETI Institute, and is also Chairman Emeritus.
EdwardsAFB

USAFB in California that is the air force's flight-test center. The base's own web site admits to testing highly unusual aircraft. The base is named after Captain Glen Edwards, a 50-mission veteran of World War II. After the war he became an Army Air Force test pilot. In 1946 he began testing Northrop's flying wing. On June 5, 1946 he was killed when the YB-49 he was co-piloting crashed.
Eighty Percent

Percentage of Americans who believe the government is hiding evidence of ETs, according to a CNN/TIME poll taken in June 1997.
Electromagnetic Wave Ducting

A process which can give false readings on a radarscope. Echoes are obtained from distant ground locations because of refraction. This has been offered as an explanation for anomalous objects appearing on radar, often the most convincing of UFO reports.

"Escalation of Hypotheses"

A phrase coined by J. Allen Hynek to describe how witnesses of UFO sightings attempt to explain and come to terms with what they have seen. Credible witnesses usually first try to fit the anomalous observation into a conventional category such as an airplane or meteor. When such explanations prove unsatisfactory, the witness then considers the extraterrestrial hypothesis. Serious UFO investigators tend to be skeptical of those witnesses who immediately assume their sighting was of an alien spacecraft.

Excelsior

A high altitude balloon project carried out in the late 1950s in order to study problems associated with the ejection of pilots from high altitude aircraft. At that time no pilot had successfully bailed out of the new jets like the F-I04 Starfighter or the U-2 spy plane. Excelsior I and II both took place in 1959. On August 16, 1960 Captain Joe Kittinger jumped 102,800 feet reaching a speed of 714 mph and becoming the first vehicle-less man to break the sound barrier. That jump also broke the record for the highest open gondola balloon ride, the longest free fall, and the longest parachute descent. Other high altitude balloon experiments also took place over the deserts of New Mexico, an area that generated a great many UFO sightings.

Exeter Incident

A classic UFO sighting that occurred on September 3, 1965 near Exeter, New Hampshire. One witness said the silent object was between 80 - 90 feet long with a belt of pulsing red lights around its middle. Another witness claimed a glowing red object had hovered over her car before shooting straight up and disappearing. For several days afterward, other witnesses reported similar sightings. Two air force officers from nearby Pease AFB interviewed three of the witnesses and other officers were seen in the area where the sightings had been reported. A colonel later said that the UFO was nothing more than the glare of landing lights at the base.

Exotic Propulsion Aircraft

Highly advanced propulsion systems that allow hypersonic flight. Exotic propulsion concepts are used in highly classified black aircraft and include pulsed detonation engines (FOEs), external combustion engines and wave riding aerodynamics or pulsed detonation wave engines.

Extremophiles

Creatures that live in the harshest, most extreme, and most toxic environments on Earth. Could this imply that life might exist elsewhere in the Universe, even on the most inhospitable planets or perhaps on a moon like Europa?

FDL-5

A research vehicle that evolved from a hypersonic research program begun in the early 1960s by Lockheed and the USAF Flight Dynamics Laboratory. The Aurora is believed to have been the final result.

Fermi Paradox

Given that planetary formation is universal, and the conditions for life and the emergence of intelligent life are the same throughout the Universe, then where is everybody? Why haven't we been contacted or visited? The Nobel Prize winning physicist Enrico Fermi is supposed to have raised this point at a social gathering at Los Alamos in 1943. His question still has not been answered.

Flying Wings

A type of aircraft developed by Reimar and Walter Horten in Germany during World War II. Jack Northrop in the U.S. also worked on the concept. These advanced, secret aircraft were unlike any other aircraft of that time. They were revolutionary in design with a delta-shape and swept bat-like wings. The flying wing lacked a fuselage and tail and could rotate around its lateral axis. They were painted with radar-absorbing paint. Northrop's XB-35 flying wing had a 172 foot wing span and was first flown in June 1946.

FOBS (Fractional Orbit Bombardment System)

Soviet missile warheads placed in low Earth orbit. After making one orbit around the Earth, they were diverted back toward a target area near the lower Volga River. Because the missiles were in violation of weapons treaties, they were given false registration

165

names. The missiles were a first-strike weapon designed to evade radar detection. Ordinary Russians who saw them thought they were UFOs. Soviet military authorities allowed the belief rather than admit to an illegal weapons system.

Foo Fighters

One of the enduring mysteries of World War II. In the closing days of the war allied pilots claimed they were being followed by strange lights and fireballs. The RAF's 115 Squadron reported seeing strange objects while on bombing raids over Germany. Some thought the exhausted pilots were hallucinating, others thought the fireballs were objects being remotely controlled by radio signals. Later it was suggested they were UFOs. The more plausible theory is that the pilots were seeing exhaust from the new German jets like the Me262 and the Me163. The Me163 Komet was the first rocket-powered fighter and could converge on allied aircraft at speeds of 600 mph or even 800mph. Its climb rate was more than double that of any other aircraft -- it could reach an allied bomber in less than six minutes.

Ghost Rockets

A wave of UFO sightings that occurred in July and August 1946, one year before the Arnold sighting. Seen mainly in northern Europe, they were described as shaped like a ball or projectile and colored bright green, white, red or yellow. They were usually seen at night traveling at a high rate of speed.

Gotha Factory

An aircraft factory in Fiedrichsrode, Germany where experimental aircraft were built during World War II. The Go-229 (also known as the Horten IX) was built here. On April 14, 1945 the U.S. Third Army captured the factory. U.S. Intelligence quickly moved in and sealed it off. Almost everything inside was shipped to the U.S. under tight security. It is believed that several experimental aircraft eventually resulted from the operation.

GPS (Global Postioning System)

A network of 12 satellites that surrounds the earth. The B-2 crew uses it to deliver bombs to a target. After the weapon is released, it continues to receive information from the satellites to guide it to the target. The system allows the B-2 to "launch and leave," the aircraft can drop the weapon and then speed away. GPS also allows the aircraft to deliver its payload in any weather.

Great Falls

A classic UFO sighting that occurred in Great Falls, Montana on August 15, 1950, also known as the Mariana Case. The principal witness was Nick Mariana who caught the objects with a 16mm. movie camera. The two objects were described as silvery bright, like polished silver. Mariana didn't see any exhaust, wings or fuselage. There was no sound except a "whooshing." The UFOs appeared to hover or stop in midair, then spun like a top as they picked up speed and disappeared.

Greenhaw, Jeff

A police chief in Alabama who in 1973 saw an alien while investigating a report of a downed UFO. He described a metallic suited creature standing in the middle of the road, and even managed to take a photo of it.

Groom Lake

A top secret USAF base in Nevada also known as Area 51, and part of the Nellis Range complex. The dry lakebed is ideal for the take-off and landing of secret aircraft. The site was selected by test pilot Tony LeVier for testing the top secret U-2 spy plane. Since then, many types of black aircraft have been developed and tested here. Officially, the base is unacknowledged. Major George Sillia, the Public Affairs Officer at Nellis AFB, said, "We can neither confirm nor deny the existence of Groom Lake itself, and if we can't confirm its existence, we certainly can't say anything about it."

Gulf Breeze

A city in Florida that is supposedly a UFO "hotspot."

Haines, Gerald K.

An historian at the National Reconnaissance Office that monitors and interprets data from reconnaissance satellites. His article, "The CIA's Role in the Study of UFOs, 1947-90" was first published in the spring issue of the CIA journal *Studies of Intelligence.* An unclassified version is available on the Internet.

HALO (High Altitude Low Observable)

Described as a small, silvery triangle, this aircraft is believed to be a top secret, unmanned British stealth fighter. However, Britain's Ministry of Defense denies the existence of any stealth programs in the UK. HALO is purportedly controlled by British Aerospace (BAe).

Halt, Charles

The USAF Deputy Base Commander at the time of the Rendlesham Forest Incident. The Halt memo is reprinted on the last page of this work.

Harvey

The code-name for a DARPA study with the goal of developing an aircraft with the lowest possible RCS. Several companies competed with each other to develop the aircraft. The winner was Lockheed's "hopeless diamond." The code word came from the invisible rabbit in the Jimmy Stewart movie.

Have Blue

A top secret program begun by a Lockheed team in the mid-1970s to develop a fighter aircraft with the lowest possible RCS. The result was a. diamond-shaped aircraft. It was designed to fly behind enemy lines, strike a target and zip out without being detected. The aircraft led to the development of the F-117 stealth fighter.

Hill, Barney and Betty

A couple from New Hampshire who claimed they were abducted by a UFO on September 19, 1961. Their story began the alien abduction craze. Betty Hill, known as the "first lady of UFOs," and the "grandmother of all abductions," died in 2004 at age 85.

Hill-Norton, Lord

The highest ranking military figure to go on the record about the Rendlesham Forest UFO incident. The former Admiral of the Fleet believes the witnesses saw something land because they were military people, trained to observe. He believes the Americans are in contact with UFOs and have exchanged technology with them.

Horten flying wings

German secret aircraft built during World War II. The HoV was an experimental flying wing that was a jet-powered version of a glider. The HoIX was a flying wing fighter powered by jet engines. It had a maximum speed of over 600 mph and a ceiling of 40,000 feet.

The HoX was a wooden prototype of a supersonic delta-wing aircraft designed in 1943. The British who likely incorporated it into their stealth program confiscated it.

Horten, Reirnar

German aircraft designer born in Bonn on March 12, 1915. His advanced ideas and designs resulted in the flying wings, the first stealthy aircraft. After the war, he escaped to Argentina where he became instrumental in that country's aircraft industry. He . reportedly participated in the development of the American B-47;" B-58, and B-2. He died in Argentina in 1993.

HSCT

High Speed Civilian Transport. A supersonic passenger plane with a design similar to the Concorde. It will fly at Mach 2.4 and have a range of 5,700 miles. The project began in 1990 and hopes to be operational by 2015.

Hudson Valley UFO Flap

A UFO flap that occurred in New York's Hudson Valley from 1982 to 1983. Hundreds of witnesses at various times and places saw strange looking, large, triangular objects with unusual lighting pattern. The UFOs either made a low humming sound or appeared silently. They were also seen hovering. Several ufologists including Philip Imbrogno and J. Allen Hynek were convinced these were extraterrestrial spacecraft.

Hybrid Aircraft

A lighter-than-air (LTA) vehicle that is part airship and part aircraft. Such an aircraft would be quite large and able to hover as well as move slowly and almost silently. Reports of very large triangular aircraft with unusual lighting patterns on their undersides could be helium-filled LTAs. Experimental designs and test flights are believed to go back to the 1960s. The ability to fly silently could be effected by using ducted turbo-fans.

Hynek, Dr. J. Allen

Dubbed "Mr. UFO" by the press, he was the scientific consultant for Projects Sign, Grudge and Blue Book although he claimed he essentially played no part in Project Grudge. He began his career as a confirmed UFO skeptic, but his attitude later changed. He wrote *The UFO Experience* (1972) and founded CUFOS (Center for UFO Studies). He also coined the term "Close Encounters" to classify UFO sightings. On November 27, 1978 Dr. Hynek addressed the United Nations on the topic of UFOs. He died in 1986.

Hypersonic

The ability of an aircraft to fly in the high Mach range (Mach 5-7, perhaps as high as Mach 9 or 10). An aircraft flying at such a speed would become very hot as the air in front of the plane stagnated and became trapped. Such aircraft would appear to change colors to any observer witnessing them. These aircraft would also require a huge fuel capacity because such speeds consume fuel at an extremely high rate. Specialized bases would also be needed. Lockheed, McDonnell Douglas and the USAF were all involved in hypersonic research for some 30 years. The Hyper-X was one project to develop hypersonic propulsion technologies. This project included a series of small, unpiloted, experimental vehicles with speeds above Mach 5.

Interstate 375

A highway in Nevada close to Area 51 and Groom Lake AFB. It is nicknamed the "Extraterrestrial or ET Highway".

Jacobs, Robert

On September 15, 1964 as a member of the 1369th Photographic Squadron he filmed a UFO while tracking an Atlas F missile. Two days later Lieutenant Jacobs was summoned to the office of Florenz Mansmann, the Chief Science Officer at Vandenberg AFB. Also present at the meeting were two CIA agents who left with the film which then disappeared. Mansmann told Jacobs to keep silent about the affair. The UFO was described as a saucer-shaped light that emitted brilliant, strobe-like flashes around the missile.

Johnson, Clarence "Kelly"

The "Legend of Lockheed." Kelly Johnson was the genius behind many deep black aircraft projects. For over 30 years he ran Lockheed's Advanced Development Company known as the "Skunk Works." He was the chief engineer for the F-104 Starfighter, the first aircraft to fly at Mach 2. In 1954 the CIA named him to build the U-2 spy plane. His team also built the SR-71 Blackbird.

Jung, Carl Gustav

A psychologist (1875-1961) who believed that UFO experiences were psychological in nature. In 1959 he wrote *Flying Saucers: A Modem Myth of Things Seen in the Skies.* Jung raised the question "Why should it be more desirable for saucers to exist than not?" He believed that saucer sightings reflected anxieties about nuclear weapons, and that in our scientific and technological age seeing flying saucers was more acceptable than having religious visions.

Kecksburg

A small town in western Pennsylvania that had a UFO experience on the night of December 9, 1965. Resident Bill Bulebush saw "this ball of fire coming from the northwest and headed toward the mountain, and it seemed like it was controlled. I watched it make a turn and swoop down in the trees." Grabbing a flashlight, he walked down into the ravine. He said the downed object looked like an acorn and was a "real dark red color like a burnt orange." He also thought the object had unusual writing, "like backwards writing," on it.

Keyhoe, Donald

Born in Iowa in 1897, he graduated from the U.S. Naval Academy at Annapolis, class of 1920. Commissioned as a second lieutenant in the Marine Corps, he became a naval aviator, flying both airplanes and balloons. He was appointed Chief of Information for the Aeronautics Branch of the U.S. Department of Commerce (later the FAA). During World War II he returned to active duty with the rank of major and served in the Pentagon in the Naval Aviation Training Division. After the war, he test flew a wide variety of aircraft. Initially skeptical of flying saucers, he changed his mind after interviewing numerous fliers and military officers as well as other expert witnesses. In the January 1950 issue of *True* magazine his article "Flying Saucers are Real" caused a sensation and became one of the most popular articles in magazine publishing history. He also wrote several books on flying saucers. In 1957 he became Director of the National Investigation Commission on Aerial Phenomena (NICAP). By 1973 he was convinced the CIA was mainly responsible for the UFO cover-up. He died in 1988.

landing lights

Black aircraft may turn on their landing lights while still at a high altitude. Baffled witnesses see a red light, white light, and amber lights streaking across the sky. These lights are at the aircraft's nose, aft (main gear doors), and in the middle (wingtips). Witnesses may also hear a low-pitched rumble or beating noise as the aircraft begins its landing approach.

laser (Light Amplification by Stimulated Emission of Radiation)

UFO witnesses often report intense red and/or green light emanating from the UFO. This is probably laser light. Light from a green laser can shine for a distance of 25,000 feet, almost five miles. Violet is another color, although less common than red or green. Evidence suggests secret aircraft use laser light to illuminate enemy targets for directing precision guided bombs. Lasers may also be used to destroy enemy targets.

Lawler, Bill

The Deputy General Manger, Military Aircraft Systems Division at Northrop Grumman. He believed the B-2 as part of a nuclear triad helped to end the Cold War by making Soviet air defenses obsolete.

LeVier, Tony

Lockheed test pilot who tested the first U-2 prototype on August I, 1955. He was just supposed to taxi the plane, but when he got it up to 70 knots the aircraft took off without LeVier realizing it. He chose Groom Lake for Lockheed and the CIA as a secure test site for the spy plane. Along with other test pilots, including Francis Gary Powers, he resigned from the air force to fly for the CIA.

Lippisch, Alexander

German aircraft designer (1893-1976) who worked for the Messerschmidt company. In 1943 he designed a bomber that was a delta wing with a straight trailing edge. Its proposed top speed was 646 mph. Only the center section was complete when the Americans captured it along with other Lippisch designs.

Lorenzen, Jim and Coral

Co-founders of the Aerial Phenomena Research Organization (APRO) founded in 1952. For almost 40 years APRO collected UFO data from all over the world. Their investigative network had representatives in more than 50 countries, and was especially active in

South America. In the 1970s the Lorenzens began microfilming their sighting files and three separate rolls are believed to exist. Where are they? The organization disbanded after the death of Jim in 1986 and Coral in 1988. In its heyday, APRO was very active and influential and was mentioned by the Robertson Panel. In 1966 Signet Books published *Flying Saucers; The Startling Evidence of the Invasion from Outer Space* by Coral Lorenzen. Jim Lorenzen was an electrical engineer at Kitt Peak Observatory.

M-17 "Mystic"

Myasishchevl Molniya M-17 "Mystic" was a Soviet high altitude aircraft with a dorsal gun turret that was used to shoot down secret U.S. reconnaissance balloons that floated over the Soviet Union. These top-secret spy balloons were mainly used to monitor Soviet nuclear tests. (See Project Mogul)

Mach number

The ratio of an aircraft's speed to the speed of sound in the surrounding atmosphere. Mach I is the speed of sound that varies somewhat by altitude. It is about 750 mph at sea level. An aircraft flying at Mach 2 is flying at twice the speed of sound. The faster an aircraft travels the greater its heat build-up, and the stranger it appears to eyewitnesses.

Mach 8 thermal barrier

The designation for the enormous heat build-up associated with hypersonic flight. Temperatures are in excess of 1800 degrees F

Macrahanish

A secret RAF base on the West Coast of Scotland once used as a forward base for the U.S. SEALS. In the mid-1980s it was used for testing the F-117 stealth fighter. The U-2 and SR-71 Blackbird spy planes also operated from here. The base is now supposedly closed, but it is believed to be a secret base used for hypersonic aircraft like the Aurora.

Majestic 12

U.S. government documents purportedly proving the existence of ETs and UFOs, and the role of the government in the subsequent cover-ups. It was later proven the documents were a hoax.

Malian, Lloyd

The author of an interesting article, "What Air Force Secrecy?" written for *True* Magazine's Special Report on Flying Saucers, 1967. Malian claimed he too once believed the air force was "knee-deep in a conspiracy" to hide the truth about UFOs. But after going through the files at Blue Book Headquarters, he was convinced "they were hiding nothing," and there was "no cellar under the offices of Project Blue Book" (where super-secret files were supposedly kept). While acknowledging that some UFO photos are "truly mystifying and defy precise scientific analysis," he concluded that because of their "vagueness" they were essentially worthless.

Manchester Airport

Known as the "Manchester Near Miss," this famous UFO sighting took place on January 6, 1995 in Manchester, England. A Boeing 737 was approaching the airport from the southeast at an altitude of 4,000 feet. Two British Airway pilots reported that a black, lighted, unidentified craft had passed them at high speed. The pilots said they ducked as it zipped by.

Manhattan Project

America's first deep black project that was kept so secret not even Vice President Truman knew about it. This two billion dollar effort resulted in the world's first atomic bomb. The project was conducted by a group of American and European-refugee scientists under the direction of 1. Robert Oppenheimer.

The scientists worked in deepest secrecy in Los Alamos, New Mexico. The first test explosion was on July 16, 1945 near Alamagordo, New Mexico. On August 6, 1945 a uranium bomb was dropped on Hiroshima, Japan, and on August 9 a plutonium bomb was dropped on Nagasaki.

McDonald, James E.

Professor of Atmospheric Sciences at the University of Arizona. From 1954 to 1956 he was the Associate Director of the Institute of Atmospheric Physics and was the Scientific Director from 1956 to 1957. He also served on the panels of the National Academy of Sciences and the National Science Foundation. Beginning in 1956 he became interested in UFOs and began interviewing witnesses. On July 29, 1968 he presented a "Statement on UFOs Submitted to the House Committee on Science and Astronautics" at the Symposium on

Unidentified Flying Objects in Washington, DC. Although he briefly considered the possibility that UFOs might be "semi-secret advanced technology," he concluded "UFOs are probably extraterrestrial devices engaged in something that might very tentatively be termed surveillance." He was highly critical of the Condon Report, claiming the committee spent too much time on trivial cases and not enough time on the truly puzzling cases. McDonald was particularly impressed by the similar accounts witnesses gave him: " ... a domed disk with no resemblance to any known aircraft that took off without a sound, and was gone from sight in five seconds ... " He died in 1971.

McMinnville, Oregon

A classic UFO case. On May 11, 1950 farmer Paul Trent took two clear photos of a saucer-like craft. The object was described as a silvery bright disk mixed with bronze that flew silently. A local newspaper printed the Trent photos that became among the most famous of all UFO photos. Even the Condon committee was impressed, and concluded the photos were authentic. It was later revealed the Trents as well as other residents had seen UFOs before.

ME-262 (Messerschmitt 262)

The world's first operational jet aircraft. Its first flight was in Nazi Germany on July 18, 1942 by Fritz Wendel, a test pilot for Messerschmitt. The flight only lasted 12 minutes, but it flew over 500 mph and broke the world speed record. On November 26, 1943 it was demonstrated for Hitler who wanted it built as a bomber. However, this proved to be too difficult and the aircraft was revamped as a fighter. Although 1,294 were built, only about one fourth actually engaged the enemy, yet these were very impressive in aerial combat. Guido Mutke, a former Luftwaffe pilot, claims he broke the sound barrier on April 9, 1945 while flying alone over Austria. Reaching a speed of 690mph, he claimed he broke the sound barrier when he felt his aircraft buffeting or shaking. Both of the engines lost function and the rivets flew out of his aircraft's wings. If his claims are true, this means he broke the sound barrier two years before Chuck Yeager who broke it officially on October 14, 1947 in his Bell X-I rocket plane.Mutke thinks other Luftwaffe pilots also broke the sound barrier because of several unexplained Me-262 crashes. However, his claims have not been verified.

Messerschmitt. Willi E.

German aeronautical engineer born in Frankfurt in 1898. He designed the BF-I 09, an advanced fighter used by the *Luftwaffe* in World War II. His advanced concepts also resulted in the ME-262 and the ME-163; the first rocket powered aircraft. He was also involved in sailplanes.

MIB (Men in Black)

Not the Tommy Lee Jones and Will Smith movies. MIB are government agents who supposedly visit witnesses after they report a UFO experience. Conspiracy buffs believe they belong to a unit of the "Air Force Special Activities Center." Air force personnel in plainclothes as well as FBI agents reportedly visited the Trents of the famous McMinnville, Oregon case. An air force agent in plainclothes demanded a reporter turn over the photos Mr. Trent had taken of the UFO.

In the Kecksurg, Pennsylvania incident a group of military people visited a local farmhouse belonging to the Hays family shortly after the fireball landed in the wooded ravine behind their farmhouse. Other witnesses were then afraid to come forward. A similar organization is the OSI (Office of Special Investigation). These MIB are supposed to be a police force in civilian clothes who use intimidation, chemicals, hypnosis, and trickery to get witnesses to deny what they've seen and sign false documents. They also spread disinformation by giving people injections and getting them to tell bizarre stories that destroy their credibility. Whether MIB are real or not, UFO witnesses are often genuinely afraid to come forward.

Miethe, Walter

Dr. Miethe was the director of the German circular wing aircraft program located near Prague during World War II. He also reportedly worked on von Braun's rocket projects, and then later for the U.S. after the war.

Mothership

Reportedly a large, agile aircraft with a wingspan of 150 feet. A total of five separate sightings of this aircraft have been reported since 1990. It is a delta-shaped aircraft with a light colored top and bottom with dark leading and trailing edges. It is called

"Mothership" because it's believed to be a vehicle for launching a smaller, hypersonic aerospace plane and/or a UAV.

Nellis Range Complex

A top secret military/government facility in Nevada incorporating Groom Lake and Area 51.

Northrop's Advanced Manned Concept

A classified supersonic stealth demonstrator. Designed in 1984, it is believed to be a next generation stealth aircraft. It has a rounded triangular or "manta" shape, and reportedly first flew in late 1995 or early 1996.

O'Brien Committee

A scientific committee formed at the request of the USAF to review Project Blue Book. It was chaired by Dr. Brian O'Brien, and was known as the "Ad Hoc Committee to Review Project Blue Book." In March 1966 it produced a short report of its findings. Dr. Carl Sagan was a member. The committee concluded, "that in nineteen years since the first UFO was sighted there has been no evidence that unidentified flying objects are a threat to our national security." It also said that much of the UFO data was too meager or indefinite for scientific analysis.

Operation Overcast

An allied intelligence operation that rounded up German scientists and specialists after the defeat of Nazi Germany. Many of these people had been involved with the "wonder weapons" programs and were fairly well known like Dr. Wernher von Braun. But a so-called "black list" included lesser-known individuals who were involved in many top-secret projects. The purpose of the operation was threefold: to use these people in the war against Japan, to keep them from working to rearm Germany, and to keep them out of Russian hands. Its code-name was later changed to Paperclip. When the Americans realized the full value of these people, Paperclip allowed the U.S. to grant them full immigration status.

Orient Express   Term used for a commercial aircraft that could take off from Dulles Airport, accelerate into hypersonic speed, attain low earth orbit, and then land in Tokyo in two hours.

Overberg

A secret stealth testing facility in South Africa. Countries like Germany, Russia, China and the UK have purportedly tested top secret aircraft here (presumably not all at the same time). UFO sightings reported from South Africa are believed to be linked to   ; Overberg.

Overholser, Denys

The former Director of Stealth Technology at the Skunk Works. His team designed the F-117.

Oxcart

Code-name for a top secret CIA project to build a follow up aircraft for the U-2 spy plane. The project began at the Skunk Works on January 30, 1960 after winning a secret design competition sponsored by the CIA. The final result was the SR-71 Blackbird.

Papoose

A facility supposedly found inside Area 51 at a base designated as S4. The base is purported to be a secret flying saucer engineering and testing facility where alien technologies are studied.

Phantom Works

The advanced R&D unit of the Boeing Company, similar to Lockheed's Skunk Works.

Pope, Nick

From 1991 to 1994 he investigated UFO reports for the British Ministry of Defense (MoD). His main responsibility was to determine if UFOs presented any threat to the national security of the UK. He concluded that UFOs were real, extraterrestrial in origin and had potential significance for national security. His book, *Open Skies, Closed Minds,* was a sensation, partly because it showed the UK had been conducting official UFO investigations and here was the guy in charge.

Predator

An unmanned drone used in Afghanistan in the war on terrorism. It is the first armed UAV (UCAV) and fires Hellfire anti-tank missiles. The total system includes four air vehicles, a ground control station, and a primary satellite link. It takes 55 people to operate it. It is equipped for reconnaissance, surveillance, target-finding missions, and now combat missions. It can hover for 24 hours, but because of its slow speed, it is vulnerable to missiles and anti-aircraft fire. Before the war on terrorism, Predators were deployed in Iraq, Bosnia, Kosovo, and in Afghanistan (before September 11, 200 I) to aid the anti-Taliban forces. Prior to September 11 drones had bin Laden in their sights but could not destroy him because they were not yet armed. Now the Predator and Global Hawk can work together as a fast reaction strike system that can hit targets almost as soon as they are detected.

Project Blue Book

The USAF investigation into UFOs from 1952 to 1969 sometimes unlovingly called "The Society for the Explanation of the Uninvestigated." A total of 12,618 sightings were investigated with 701 remaining unidentified. When Blue Book closed the USAF was officially out of the UFO business, but critics all along had charged that it had never mounted a thorough scientific investigation of UFOs.

Project Mogul

A super secret operation begun in 1947 that involved the use of reconnaissance balloons which floated over the Soviet Union in order to monitor Russian efforts to build and test nuclear weapons. Mogul was so secret that its national security rating was "Top Secret A-I," the same rating as the Manhattan Project. It was a top secret Mogul balloon that crashed in Roswell, New Mexico. The Soviets were fully aware of these balloons and had developed a high altitude aircraft to shoot them down, the M-17 "Mystic."

Project Ozma

The first attempt to detect radio signals from other worlds. It was conducted from April to July in 1960 by radioastronomer Frank Drake. The search was aimed at two stars, Tau Ceti and Epsilon Eridani, both about eleven light years from earth.

Pulsed Detonation Engine (PDE)

A hypersonic propulsion technique also known as a pulsed detonation wave engine. It uses a shock wave created by a detonation, an explosion that grows supersonically. The aircraft in effect rides its own shock wave. This technique allows aircraft to fly in the high Mach range, but it also generates tremendous heat build-up. PDE engines can perform the "hyperjump" maneuver: an aircraft equipped with such an engine would accelerate at a speed greater than the human eye could follow. This would make the aircraft seem to

179

disappear to the observer.
RADAR
Radio detection and ranging. Radar is a system for detecting the position, motion, and nature of an object by reflecting radio waves from its surface. The cavity magnetron is the innovation at the heart of radar. It was developed by the British during World War II and helped them win the Battle of Britain. Radar Cross Section (RCS) is the size a radar system indicates an object to be. The lower an aircraft's RCS, the less visible it is to radar systems. The B-2's RCS is about the size of a bee.

RAF Boscombe Down
The headquarters of the UK's Defense Testing and Evaluation Organization (DTEO) which is part of the Defense Evaluation and Research Agency (DERA). It has been a flight testing center since 1914, and it is believed secret aircraft are being tested here today.

Ramjets
These engines rely on their forward motion to compress air. The air is scooped up and compressed as it funnels into a diffuser. The diffuser acts to slow the air for easier combustion. The air and fuel are transferred to a chamber for combustion. An exhaust nozzle accelerates the burst after ignition producing the thrust.

Reagan, Ronald
In 1988, while meeting with the Soviet leader Mikhail Gorbachev, President Reagan remarked: "With our obsession with antagonisms of the moment, we often forget how much unites all the members of humanity. I occasionally think how quickly our differences worldwide would vanish if we were facing an alien threat from outside this world." In 1999 President Clinton echoed Reagan's comment when he said: "If we were attacked by Space Aliens we wouldn't be playing these kinds of games." But perhaps the most provocative quote came from General Douglas MacArthur in 1955: "The nations of the world will have to unite, for the next war will be an interplanetary war. The nations of the Earth must someday make a common front against attack by people from other planets."

There is speculation that fear of an alien attack was President's

Reagan's rationale for pushing the SDI (Star Wars) program (see entry). Suspicion increased when SDI was revived under President Bush II; Defense Secretary Donald Rumsfeld said that an anti-ballistic missile defense system was needed to protect the U.S. from possible attacks by terrorists or rogue states. Critics immediately pointed out that terrorists would most likely not use missiles, but would rely instead on low-level technologies. Why develop a billion/trillion dollar system to take out a guy who rides a camel and lives in a cave? Why would we need SDI, except in the case of a *War of the Worlds* scenario?

Roswell

A small town in New Mexico where a UFO was thought to have crash-landed on a nearby ranch in 1947. The owner of the ranch, "Mac" Brazel, reported the crash to the local sheriff who in turn contacted the military authorities at nearby Roswell Army Air Field. Major Jesse Marcel was assigned to the investigation. On July 8, 1947 the public information office at Roswell AAF announced that the remains of a flying disc had been recovered. However, on the following day, Brigadier General Roger Ramey along with Major Marcel held a press conference and announced the remains belonged to a weather balloon. This explanation did not satisfy UFO buffs, and over the years, Roswell became the premier UFO event. It is now known that what really crashed was a top secret spy balloon.

Rowehl Geschwader

The first reconnaissance aircraft, supposedly, capable of flying at high altitudes and at high speeds. It was developed in Germany during WW II for spying over the Soviet Union. The top-secret aircraft had pressurized cabins, engines specially developed for high-altitude flying, and special photographic equipment with a wide angle of vision. The idea was to fly so high (33,000-39,000 feet) the Soviets wouldn't notice anything. After the war, the Americans obtained possession of the program's secret documents and these likely aided in the development of the U-2 spy plane.

Ruppelt, Edward J.

As Director of Project Blue Book from 1951 to 1953, he created the term "UFO" to replace flying saucer. In 1956 he wrote *The Report on Unidentified Flying Objects,* which contains some of the early, classic UFO sightings. Ruppelt reportedly confided to Donald Keyhoe that the CIA was behind the air force's debunking of UFOs. "They ridicule witnesses, discredit our own pilots, and hide sightings." His remarks drew a heated denial from the air force.

Russell, Richard

A Georgia senator who went on a fact-finding trip to the Soviet Union in October 1955. While traveling by train through southern Russia, he saw two disk-shaped aircraft take off vertically, rise slowly and then shoot off at high speed. Both the CIA and FBI conducted investigations and concluded the senator had really seen Soviet jet fighters. Russell's report was said to be buried in secret files.

Sagan, Carl

One of the world's most influential scientists, especially in the field of space exploration and the search for extraterrestrial life. Dr. Sagan was the Director of the Laboratory for Planetary Studies and David Duncan Professor of Astronomy and Space Sciences at Cornell University. He played a major role in the Mariner, Viking, and Voyager unmanned space expeditions. He wrote numerous books and articles including the novel *Contact* (with wife Ann Druyan) which was made into a movie. He also won many awards including the Pulitzer Prize, but he is probably best remembered for *Cosmos,* his popular book and 1977 PBS television series. He died in 1996. He mentions the Aurora in his book *The Demon Haunted World.*

Saunders, David R .

A principal investigator on the Colorado Project who began to computerize the data on UFOs. He had several thousand cases on magnetic tape. Dr. Saunders, along with Dr. Norman Levine, was dismissed from the project. Their dismissal, in turn, lead to the resignation of Mary Louise Armstrong, Condon's administrative assistant. In 1968 Saunders, along with Roger Harkins, wrote *UFOs? Yes!* in which they blasted the Colorado Project for its "unscientific, biased, and pre-concluded findings" and they concluded UFOs were extraterrestrial spacecraft.

Scramjets (air-breathing engine)

In normal jet engines rotating blades suck in and compress the air, but scramjets have no rotating parts. Instead, they operate by the supersonic combustion of fuel in a stream of air, compressed by the forward speed of the aircraft. The supersonic air is compressed through a narrow opening by shock waves to a pressure greater than the surrounding air. The fuel is fanned by the supersonic airflow and ignites. The expanding hot gas increases thrust, and the aircraft is propelled forward by the hot exhaust.

SDI (Star Wars)

The Strategic Defense Initiative (SDIl), first proposed by President Ronald Reagan in 1983, was a military research program aimed at developing an anti-ballistic missile (ABM) defense system. SDI involved highly advanced weapons technologies with many of them only in their research stage. It included such *Star Wars* weapons as space and ground based nuclear X-ray lasers, subatomic particle beams, and computer-guided projectiles fired by electromagnetic rail guns, all of which was supposed to be controlled by a supercomputer system. Another part of the system included a network of space-based sensors and specialized mirrors for directing the laser beams toward targets. Funding such a system would cost in the billions, if not trillions of dollars. Highly controversial, SDI was abandoned in 1993 only to be revived under President Bush II. Some believe the real purpose of SDI is to develop a defense system against possible alien invasion (See Ronald Reagan).

Senior Citizen

An advanced tactical low-observable aircraft. It purportedly was funded until 1993 when it disappeared from the budget, suggesting it was either terminated or had entered active service. It might be the alternate designation for Aurora or it could be a similar class of aircraft.

SERENDIP

The Search for Extraterrestrial Radio Emission from Nearby Developed Intelligent Populations. This is an ongoing SETI project at the University of California at Berkeley. SETI is a privately funded Search for Extraterrestrial Intelligence that uses giant radio telescopes to scan outer space for radio signals from other worlds. So far SETI has not received a confirmed, extraterrestrial signal. The SETI

Institute runs Project Phoenix, a targeted search that focuses on nearby sun-like stars. The receivers look for constant or slowly pulsed narrow-band "carrier" signals. Any signal less than 300Hz wide would probably be artificially produced. In 1977 the Ohio State Radio Observatory picked up the "Wow" signal, but it was only detected once and was not repeated. If an ET signal is ever detected, the SETI Institute would not reply. Instead, the nations of Earth would make a decision on whether or not to reply and what kind of reply to send. SETI does no UFO research.

Seymour, Scott

Scott was the B-2 Program Manager at Northrop Grumman.

SHAFFT

A supersonic/hypersonic fighter aircraft that is to be part of an integrated weapons system for the 21st century. The fighter would be triangular in shape and use the PDE wave rider principle. Its purpose: fly deep into enemy territory and strike a lethal blow and then zip out without being detected.

Sheraz, Iran

In June 1978 a 16-year-old Iranian student photographed a UFO from his family's apartment window in the town of Sheraz. The object resembled a UFO that had been earlier reported by pilots landing at the Teheran airport. In October, a similar object was photographed in western Iran. Ufologists believe Iran long has been a UFO hotspot, especially in September 1978 when two Iranian jets encountered an object that did not appear on radar.

Silver Bug

A circular-wing aircraft that was supposedly built as a weapons and/or reconnaissance platform. Both the U.S. and UK were involved in its development.

Smart Skin

Highly advanced stealthy technologies that allow an aircraft to change color to match its surroundings. It involves the use of sensors and computers. Prototypes have been reportedly tested in the UK.

Socorro

On April 24, 1964 at 5:45 P.M. a famous Close Encounter took place near Socorro, a small town in New Mexico. Police Officer Lonnie Zamora was chasing a speeding car when he suddenly heard a loud earsplitting roar. He then spotted a "flame in the sky" near a spot where there was a small shack for storing dynamite. Fearing the dynamite had exploded, he gave up on the speeding car and headed for the shack which was on top of a mesa. When he got there, he saw a shiny object with "two people in white coveralls" beside it. At first sight, the object looked like "a car turned upside down," then later like an egg-shaped, metallic craft. A strange insignia was on the object's side. As Zamora got out of his car, he heard "a very loud roar." A giant flame blasted from underneath the object as it rose straight up and then took off horizontally. Dr. J. Allen Hynek, who investigated the case, tried to get the air force to participate, but to no avail: "Because it was merely a UFO case, the usual pattern of doing nothing was followed."

Sound Barrier

The boundary between subsonic and supersonic flight. Flying in his X-I rocket plane at 670 mph, it was first breached by Chuck Yeager in 1947. A sonic boom is created when an aircraft roars past Mach I and emits intense shock waves.

Space Shuttle

Built in the 1970s, the vehicle is used to transport astronauts into space. It takes off like a rocket at 17,500-mph using three liquid rocket boosters, but lands like an airplane. Conspiracy theorists believe the "white" shuttle program is used to hide "black" funding for top secret projects.

Special Relationship

The close partnership between the U.S. and UK which began during World War II. The relationship exists on a number of levels including language, culture, economic, political, and especially military. It is believed the two countries have cooperated closely on black projects like stealth. The relationship also involves close cooperation between the CIA and its British counterpart M16. It also allows for the U.S. to operate spy planes like the U-2 and SR-71 from the UK. The Aurora and possibly other deep black aircraft can also operate from bases in the UK.

Speed of Light

186,000 miles per second (300,000 km/sec). It would take 4 years for the light from our sun to reach Alpha Centauri, our nearest neighbor. Light speed is believed to be the Universe's maximum speed limit, although this may not be true at the subatomic level. The enormous distances involved in space travel may be a major reason for the lack of contact between alien species.

"Spooky"

A special version of the AC-130 gunship that was specifically designed for the Special Forces. Its ability to fly high and fast makes it undetectable by radar. Complex electronics allow it to rapidly fire precision guided 25, 40, and 105-mm projectiles.

"Sport Model"

A UFO supposedly housed at S-4 in Area 51. Some ufologists believe black aircraft are back-engineered from captured UFOs like this one.

Sputnik

On October 4, 1957 the Soviet Union launched the first artificial satellite. It took pictures of the dark side of the moon. The launching of *Sputnik* shocked the world, especially the U. S., and the "space race" began, culminating with Neil Armstrong's moon landing on July 20, 1969.

Strangeness Spectrum

A term used by J. Allen Hynek to describe the "definite pattern of strange craft" seen in UFOs. UFOs do not run the gamut of all possible strange configurations, but rather show similar data patterns and a measure of repeatability. Craft are triangular or ovoid, can be silent, motionless or hovering, and are frequently given to great bursts of speed. Anytime "aliens" are seen they are humanoid in appearance.

Symbols

Symbols, writings, or markings sometimes reported on close encounters with UFOs. In the Roswell case, Jesse Marcel Jr. saw strange symbols on the wreckage his father Major Jesse Marcel brought home. In the Kecksburg case, witnesses saw unusual symbols on the object that crashed in the Pennsylvania woods. They said it looked Egyptian or Russian. In the Rendlesham Incident, Jim Penniston saw symbols etched or engraved on the smooth, black object that landed in the forest. Lonnie Zamora saw a strange insignia on the UFO he saw in New Mexico.

Tacit Blue

A low observable surveillance aircraft with a number of innovative stealth features. It was part of a highly classified program that ran from 1978-1985. It entered the white world because its technologies and capabilities are currently in operational use, and the program no longer needs to be protected. Tacit Blue was one prototype of stealth technologies, and provided data that was used in the B-2 design and possibly the Aurora. It was one of the most successful technology demonstrator programs in air force history. The cost of the program was $165 million.

Team Stealth

A combination of stealth aircraft, perhaps unmanned as well as manned, all working together. Team stealth is believed to include the B-2, F-117, the Black Manta, and perhaps the Commanche helicopter and the British HALO. It may also include more advanced aircraft like Aurora. It was probably a Team Stealth exercise that was responsible for the Belgian UFO flap and perhaps other UFO flaps where several unidentified objects were reportedly working in unison.

Tier I, II, III

UAV projects. Tier I was begun by the CIA in order to quickly develop a tactical reconnaissance/surveillance UAY. The eventual result was the Predator. Tier II Plus (Global Hawk) was part of a larger program aimed at developing a non-stealthy, long endurance UAY. It is called HAE (High Altitude Endurance). Tier III Minus, or DarkStar, was a stealthy reconnaissance UAY that looked like a flying saucer. Tier Plus indicates a conventional design while Tier Minus is a low observable shape.

Titanium

A material that is as strong as stainless steel, but one-half the weight. It can withstand high temperatures and pressures. Titanium was used to build the SR-71 Blackbird as well as other stealthy aircraft so these aircraft would be able to sustain the high temperatures resulting from flying at supersonic speeds.

Title 14

Title 14, Section 1211 of the Code of Federal Regulations was implemented on July 16, 1969. It makes it illegal for U.S. citizens to have any contact with ETs or their vehicles.

TSR2

One of the earliest of the flying triangles reported in the UK. It was reportedly a prototype of a secret, stealth-like aircraft that crashed near RAF Upper Heyford in 1967. The debris was then taken to RAF Bicester under great secrecy. The program was purportedly cancelled when the Wilson government deemed it too expensive. However, there is ample evidence that Britain does have a stealth program.

UAS, UAV

Refer to the beginning of chapter eleven

Two-stage-to-Orbit

An aircraft/spacecraft system in which a high speed aircraft flying at Mach 6 - 8 releases a launch vehicle. The unmanned, stealthy, ramjet drone then accelerates to orbital velocities and releases a small satellite into space.

Ufimstev, Pytor

A Russian physicist whose formulas for computing the scatter of electromagnetic energy over a surface helped to create stealthy aircraft. The Soviet government dismissed his theories and thereby missed out on one of the greatest technological advances of the 20th century. Today he works as a consultant for Northrop.

The UFO Enigma: A New Review o/the Physical Evidence

A book by Peter A. Sturrock, the director of a scientific study of UFOs which took place at the Pocantico Conference Center in Tarrytown, N. Y. from September 29 to October 3, 1997. The book is significant because it was published in 1999 and claims to be "the first major scientific inquiry since the Condon Report." While critical of the University of Colorado Project, it makes the same mistake of failing to consider the possibility of top secret aircraft. This is especially unusual since some secret aircraft had entered the white world by 1997 and certainly by 1999.

V-2

The first modern rocket was developed by German engineers during World War II. The V-2 had two liquid boosters made of pure alcohol and oxygen and one warhead. After the war, the U.S. took these rockets to White Sands. The V-2 was developed at Peenemunde on the Baltic Coast where a device called *Wasserfall,* a ground-controlled anti-aircraft rocket, was also developed. The V-1 or "doodlebug" was a robot bomb.

Valentich, Frederick

On October 21,1978 the 20-year old Australian pilot was flying his single engine Cessna when he disappeared between Melbourne and King Island. He reported a large, bright metallic craft flying at high speed. His last words: "That strange aircraft is hovering on top of me again .. .it is hovering and it's not an aircraft." No trace of the pilot or his plane was ever found.

VTOL

Vertical Take off and Landing. Sometimes witnesses reported UFOs that seemed to sit on their tails and fly straight up. The XFY -1 ("Pogo") was a 'tail-sitter," a delta-winged aircraft that could take off vertically before making the transition to regular, horizontal flight. To land, the pilot had to pull up vertically into a hover position and then back the plane down. The X-13 Vertijet was the first jet VTOL aircraft. More recently, the U.S. navy is experimenting with a hybrid jet that would take off vertically like a tail-sitter and then land conventionally or use vectored thrust.

Warton

A facility in Lancashire belonging to British Aerospace where secret, classified aircraft are built. It is the British equivalent of the Skunk Works. Triangular UFOs are frequently spotted in the surrounding area. Stealth aircraft are thought to be built as the Special Projects Site including "next generation" stealth.

Waveriders

Wedge-shaped aircraft that surf on a stream of air Shock waves build up during high-speed flight and increase lift. Sharp leading edges on wave riders wrap the shock waves around their fuselages. A cushion of air is then trapped below it.

West Freugh

A remote RAF base on the south coast of Scotland that includes a test range for secret aircraft like HALO. Several small triangles have been seen in the area.

White Projects

Advanced aircraft programs that are acknowledged by the military/government and are openly funded and debated. The space shuttle is an example. Conspiracy buffs believe one way black projects are secretly funded is by hiding them in the budgets for white programs -- so there may actually be a reason for those $10,000 toilet seats! Sometimes black projects are eventually revealed and then enter the white world.

Wonder Weapons

Advanced weapons technologies developed in Nazi Germany during World War II. These programs included not only the V-I and V-2 rockets, but also remote-controlled gliders, advanced anti-aircraft devices, the first jets, circular-wing aircraft, the first stealth-like aircraft, and high altitude spy planes. After the war, it was discovered that Japan had also been developing highly advanced aircraft as well as the first bioweapons.

X-IS

A rocket plane that flew at an altitude of353,760 feet (67 miles) at a speed of Mach 6.7 (4,520 mph). It was dropped from a B-52 bomber. The program ended in 1968 after 199 flights by three test aircraft. The X-IS was one of the most successful experimental aircraft programs in the history of the USAF. Unlike the rocket-powered X-IS, the next generation hypersonic aircraft have air-breathing engines.

X-22A

A discoid-shaped aircraft built by Lockheed that utilizes anti-gravity propulsion. The two-man craft was believed to have been spotted during Desert Storm by both U.S. and Iraqi soldiers. The craft fires intense beams of light (probably lasers) that vaporize everything at that spot and then leaves behind a circular, charcoal-like burn mark. This craft is believed to be the real DarkStar, and the DarkStar drone was a cover for the manned anti-gravity fighter.

X-30 (NASP)

The program to develop the National Aerospace Plane (NASP) designated the X-30. It has its roots in the highly classified deep black project called Copper Canyon and was originally conceived as a single-stage-to-orbit (SSTG) airplane that could take off and land horizontally. It is believed to be a follow-on to the SR-71. Conspiracy buffs suspect the budget for the X-30 also hides the Aurora's big budget.

X-36

A prototype tailless fighter aircraft designed for stealth and agility. Developed by Boeing at its Phantom Works and NASA, the tailless design reduces the aircraft's weight, drag and RCS. The flight test program began in 1997 at Edwards AFB.

X-43

A small, unmanned research scramjet that is part of NASA's Hyper-X program. The air-breathing aircraft scoops and compresses oxygen from the atmosphere using the vehicle's airframe. The scramjet is carried on a modified Pegasus booster rocket. After reaching a certain altitude and velocity, it will separate from the booster and fly under its own power as fast as 7,000 mph. Because of the severe heating associated with such high speeds the aircraft glows yellow.

The Holt Memo-13 Janary 81
Subject: Unexplained Lights
To: RAF?CC

I. Early in the morning of 27 Dec. 80 (approximately 0300L), two USAF security police patrolmen saw unusual lights outside the back gate at RAF Woodbridge. Thinking an aircraft might have crashed or been forced down, they called for permission to go outside the gate to investigate. The On-duty flight chief responded and allowed three patrolmen to proceed on foot. The individuals reported seeing a strange glowing object in the forest. The Object was described as being metallic (sic) in appearance and triangular in shape, approximately two and three meters across the base and approximately two meters high. It illuminated the entire forest with a white light. The object itself had a pulsating red light on top and a

191

bank(s) of blue lights underneath~ the object was hovering or on legs. As the patrolman approached the object, it maneuvered through the trees and disappeared. At this time the animals on a nearby farm went into a frenzy. The object was briefly sighted approximately an hour later near the back gate.

2. The next day, three depressions 1 – 1 ½ " deep and 7" in diameter were found where the object had been sighted on the ground. The following night (29 Dec 80) the area was checked for radiation. Beta/gamma readings of 0.1 milli-roentgens were recorded with peak readings in the three depressions and near the center of the triangle formed by the depressions. A nearby tree had moderate (0.5 - 0.7) readings on the side of the tree toward the depressions.

3. Later in the night a red sun-like light was seen through the trees. It moved about and pulsed. At one point it appeared to throw off glowing particles and then broke into five separate white objects and then disappeared. Immediately thereafter three star-like objects were noticed in the sky, two objects to the north and one to the south, all of which were about 10 degrees off the horizon. The objects moved rapidly in sharp angular movements and displayed red, green and blue lights. The object to the north appeared to be elliptical through an 8-12 power lends. They then turned to full circles. The objects to the north remained in the sky for an hour or more. The object to the south was visible for two or three hours and beamed down a stream of light from time to time. Numerous individuals, including the undersigned, witnessed the activities in paragraphs 2 and 3.

   (Signed) Charles I. Halt. Lt Col. USAF Deputy Base Commander

The End of This Book
For a bibliography of this research, documentation, and photos of UFOs and other scenes, the reader is directed to one of the earlier print version of this work. i.e ISBN 978 0910042 031.

   The first print version is titled *Out of The Black* and was published by Allegheny Press.

The U. S. Navy's E – 6 Mercury looks like a commercial airliner, but one can be deceived by appearances. It doesn't carry any weapons but it is probably the deadliest aircraft in existence. Its duty is to command the launch of land and sea-based nuclear missiles.

The air-borne mission of the E 6 is combined with land based transmitters in strategic locations in the U.S. If the ground based operations are eliminated the E-6 will still be in contact with the president and the secretary of defense, as well as our nuclear forces. If an enemy destroys the U.S. ground based command, this doomsday plane will still function.

www.ingramcontent.com/pod-product-compliance
Lightning Source LLC
Chambersburg PA
CBHW051501170526
45166CB00001B/340